Artificial Intelligence for Fashion

How AI is Revolutionizing the Fashion Industry

Leanne Luce

Apress®

Artificial Intelligence for Fashion: How AI is Revolutionizing the Fashion Industry

Leanne Luce
San Francisco, CA, USA

ISBN-13 (pbk): 978-1-4842-3930-8
https://doi.org/10.1007/978-1-4842-3931-5

ISBN-13 (electronic): 978-1-4842-3931-5

Library of Congress Control Number: 2018965464

Managing Director, Apress Media LLC: Welmoed Spahr
Acquisitions Editor: Natalie Pao
Development Editor: James Markham
Coordinating Editor: Jessica Vakili

Cover image generated by a generative adversarial network trained by Leanne Luce.

Distributed to the book trade worldwide by Springer Science+Business Media New York, 233 Spring Street, 6th Floor, New York, NY 10013. Phone 1-800-SPRINGER, fax (201) 348-4505, e-mail orders-ny@springer-sbm.com, or visit www.springeronline.com. Apress Media, LLC is a California LLC and the sole member (owner) is Springer Science + Business Media Finance Inc (SSBM Finance Inc). SSBM Finance Inc is a **Delaware** corporation.

For information on translations, please e-mail rights@apress.com, or visit http://www.apress.com/rights-permissions.

Apress titles may be purchased in bulk for academic, corporate, or promotional use. eBook versions and licenses are also available for most titles. For more information, reference our Print and eBook Bulk Sales web page at http://www.apress.com/bulk-sales.

Any source code or other supplementary material referenced by the author in this book is available to readers on GitHub via the book's product page, located at www.apress.com/978-1-4842-3930-8. For more detailed information, please visit http://www.apress.com/source-code.

Printed on acid-free paper

This book is dedicated to all women who defy the expectations to fit neatly into the roles society provides.

Especially to the fierce women in my life whose example shaped who I am: Evelyn, Barbara, Meredith and Jeanette.

Table of Contents

About the Author

Leanne Luce is a graduate of the Rhode Island School of Design (RISD). She has a background in fashion and previously worked on robotic exoskeletons and soft goods for the military at the Wyss Institute at Harvard University and Otherlab in San Francisco.

Leanne is also editor of the blog The Fashion Robot, which has been recognized by publications such as *Vogue* and *Glossy*. She currently works as a Product Manager at Google.

Acknowledgments

I want to thank my partner, Will, the most of all. Some days it felt as though I would not be able to finish this book without his continual support, encouragement, and knowledge. His shared passions for software and creativity have been inspirational and motivating. My gratitude and my love cannot be expressed in words.

Thank you to all of the friends and colleagues who shared their knowledge and gave their support and encouragement throughout the writing of this book including Tom Brown, Dan Corkum, Ziwei Jiang, Michael Rubin, and Carver Wilcox.

Thank you also to Palaniswamy Rajan, Mary Ouk, and Adam Bouhenguel for their expertise through parts of the book. Thank you especially to Adam for your patience, advice, and friendship.

Preface

The rest of this book won't dive into the autobiographical, but for readers who don't work in the tech industry, I want to share the meandering story about how I personally got here.

I started out my career as a fashion designer. I didn't immediately have the intention that I would work in tech, but it didn't just happen either. Working in fashion, I was interested in how and where things were made. I cared less about what or who was "cool" that season unless it helped me to solve a problem. I was frustrated with the way the fashion industry worked. It was probably not surprising to those around me when I made the decision that the industry fashion design track was simply not for me.

Early on, I started seeking opportunities to work on innovative technologies where my skills could be valuable. I found a team at the Wyss Institute at Harvard that was building wearable soft robotic exoskeletons, a project they called the Exosuit. The project proposed a number of applications from military to stroke rehabilitation. My involvement did not start out in a glamorous way. I started sewing prototypes and worked my way into other parts of the development process: observing user-testing, suggesting improvements. While I wasn't necessarily excited about sewing at the time, I was refreshed by the sense of meaning I got from building wearable devices that could change people's lives by helping them move.

Shortly after working at the Wyss Institiute, I moved to San Francisco and began working at Otherlab on a similar project. I became interested in how the technologies I was using to build these suits worked and how some of them might be used to improve processes and products in the fashion industry. Eventually, I left my job in robotics to start a company to develop an automated technique to manufacture custom-fit bras. The company failed within a year, and I started reflecting about why it had failed. I wanted to build a software system to take in user measurements and send instructions to a machine that would output a garment with minimal human intervention. However, I was missing a lot of high-level information about how I would build the software to do it.

Change

As I was exploring options for my next career move, I started writing. I wrote about a wide range of technologies and started keeping a list of software and technology companies that were entering the fashion arena. Writing also opened up the opportunity to talk to people who were working in the space, learning about what they were working on and what their personal journey had been to get there. I needed to make a decision about what I would do next. At that time, I started asking friends for advice and one piece of advice stuck with me: "Programming is only becoming more relevant every day." It didn't help that I was living in the Bay Area, but when I looked around, I could only find evidence to support that statement. The companies that I believed were having the biggest impact in fashion were almost all software companies. In my mind, acquiring programming skills would be extensible to almost anything I could imagine myself working on next, including the company I had tried to create.

I decided that I would learn to code. The intention was never that I would be the greatest engineer. I wanted to get enough under my belt that I could understand it. Once I decided I wanted to learn to code, I had a lot of work to do to get to the point that I could make that a viable goal. In the beginning, I didn't realize that the term *software* expanded into a vast technical space and that some of the branches required skills that might not be overlapping. As an outsider, I could not begin to imagine the nuances within the industry. It felt so far from my reach and far from my understanding. I questioned myself, my own intelligence, my own ability to understand.

Failure Again

I asked friends how they learned and what they recommended, many of them recommended the Introduction to Computer Science and Programming course on MIT's OpenCourseWare. I started and restarted the MIT course three times over the course of six months before I was willing to admit that it was not working for me. Plenty of people doubted me through the process, but at that point, even some of the most loving and supportive people in my life were starting to dissuade me. They suggested that maybe my inability to finish this course meant that I was just not that interested. In many ways, I took this failure streak to heart.

One of the biggest reasons I couldn't finish that course was that I didn't understand what skills I would acquire from it. I couldn't *see* the results of what I was learning—and as a visual person, I needed to. I thought that going through the course would help me to understand the bigger picture.

Finding an Entry Point

I sought a different entrance to the world of software and programming. This time, I looked for an entrance that allowed me to see the results of what I was learning immediately. I started with basic HTML and CSS, and found that was my entry point. I later understood that this was called *front-end web development*. While this all seems obvious to me now, at the time it was not. I had a very hard time making sense of it all.

Over time, that barrier to understanding shortened, and summiting it became easier and easier. There was no magical solution, just perseverance. Finding that entry point was the most important thing for my experience. Even though it gave me a shallow initial understanding, it also gave me a place to start asking questions and unravelling the large ecosystem that is software. After learning HTML and CSS, I went on to learn some Javascript and Python. Learning enabled me to think about software more wholistically and further develop ideas of my own. While I never intended to be a software engineer, for me learning to code was a crucial step to becoming a Product Manager at Google.

Learning about AI

When I started learning about artificial intelligence, I was faced with the same challenge. The topics in AI were complex concepts that I did not yet understand. This time, the problem was more complicated because it relied on foundational information that I sometimes didn't have. The resources available to learn about AI were usually limited to two generalized extremes:

1. Technical literature which focused deeply on the mathematics and implementation, diving so far into the equations that there was not enough information to understand from a high level.

2. Marketing and business resources, which provided lots of reasons to use AI, but little to no explanation of what it was or how it worked. It was not enough information to understand.

The Guide I Wish I Had

The experience I had learning about AI was frustrating. In writing this book, I set out to create a resource that would take away some of that frustration for other curious people, especially in the fashion industry. This book is meant to help people find an entry point and gain the high-level understanding that I was seeking in my own research. It's the guide I wish I had when I was learning. Although the lens here is specifically focused on the fashion industry, I believe the content of this book is valuable to anyone just starting to learn about artificial intelligence. The examples used in the fashion industry are widely applicable in other industries as well.

The chapters are designed to answer some of the basic questions that come up as you unravel the world of artificial intelligence. This book is just the beginning; there will always be more to learn.

Introduction

Artificial intelligence (AI) is becoming a part of the way we conduct business in every industry. The fashion industry is no exception. From product discovery to robotic manufacturing, AI has made its way into almost every segment of the fashion value chain.

The goal of this book is not just about telling businesses and brands how to incorporate AI into their daily practices, but to help readers add AI to their toolkits.

About This Book

Artificial intelligence (AI) is a very large field. This book does not attempt to cover every topic, but to give a foundation for understanding.

The structure of this book starts with a brief introduction to artificial intelligence. After the intro, the book is broken into sections, working from consumer-facing products through to manufacturing. To give context for these applications, each chapter goes through customer or industry pain points and explores real-world examples of how fashion companies are using AI to solve these pain-point problems.

Each chapter is organized with an explanation of a major industry-based application and an explanation of a key technology concept. The chapters each reveal a little bit more about the inner workings of artificial intelligence and are cumulative.

In each section, you almost step through the departments of a fashion business and see how AI is being used in unique ways in every department. There are some similarities with using AI as there were when businesses started to use computers. At first, people thought computers would be used only in math departments, but today they are ubiquitous. You can hardly think of a field without a computer. Similarly, you might assume AI is being used only in certain industries, but AI is pervasive across industries.

The book is as direct as possible; it gets to the point with explanations about key concepts in artificial intelligence and breaks down technical jargon needed to understand these key concepts.

The Audience

This book is written so that anyone who picks it up can understand it. It is not a book for engineers trying to get better at coding or learn new techniques. There are no equations, algorithms, or code in this book. From a technical perspective, topics discussed in this book are highly simplified.

Not just one department or IT person should read this book. Adapting AI in companies affects everyone, from designers to C-level executives. Managers and employees in every department can learn how to implement new technology in their respective expertise to improve processes. It is written for the following:

- **Fashion industry professionals** who have no experience with software or coding, but are curious and interested in how technology is changing the fashion industry.

- **Executives and managers** making decisions about technology implementation at fashion brands.

- **Entrepreneurs** looking to create technology for the fashion industry.

- **Students** thinking about future career choices in fashion and technology.

PART I

Introduction

CHAPTER 1

Basics of Artificial Intelligence

Fashion not only provides functional purpose, but captures mysterious and elusive aspects of being human. Fashion expresses and invokes human emotion and creativity. How we look and sometimes even how we feel is intertwined in this industry. Fashion has always been forward looking, grabbing onto new technologies as they arise. Artificial intelligence is no exception, and it's moving as quickly as fashion does.

Artificial intelligence (**AI**) is a field of computer science that looks at the logic behind human intelligence. The field seeks ways to understand how we think and to re-create this intelligence in machines. Because of its nature, AI extends across human activities, making it relevant in different ways to every industry.

The intersection of fashion and AI is a rich and expansive space that is just beginning to be explored. As AI continues to develop, it becomes harder to comprehend for nontechnical followers. The challenge of comprehension stands in the way of meaningful developments between these two fields.

This chapter briefly covers basic concepts in artificial intelligence to provide a foundation for understanding its applications in the fashion industry. The rest of the book expands on these ideas and more.

© Leanne Luce 2019
L. Luce, *Artificial Intelligence for Fashion*,
https://doi.org/10.1007/978-1-4842-3931-5_1

Why Does AI Matter?

In "The State of Fashion 2018," a report by McKinsey & Company and The Business of Fashion, 75% of retailers plan to invest in artificial intelligence over 2018 and 2019. It is changing the way the fashion industry does business across the entire fashion value chain. Providing customized experiences and better forecasting is just the start.

Currently, up to 30% of activities in 60% of occupations across all industries can be automated. It will still take time to implement some of this automation and reskill the current workforce. At this rate, there is no question that artificial intelligence will significantly impact the way we work.

What Is AI?

Artificial intelligence has become a confusing term. Machine learning, deep learning, and artificial intelligence are terms often used interchangeably, which may leave to question, what is the difference?

Machine learning is a way of achieving AI. In 1959, it was defined by Arthur Samuel as "the ability to learn without being explicitly programmed." Usually this is done through "training." **Deep learning** is an approach to machine learning, which usually involves large neural networks. Figure 1-1 shows a graphical representation of the relationship between AI, machine learning, and deep learning.

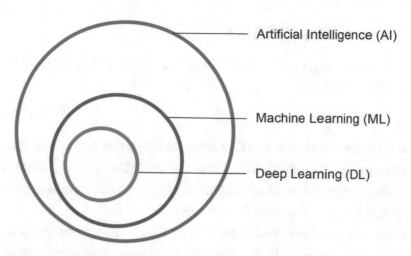

Artificial Intelligence (AI)

Machine Learning (ML)

Deep Learning (DL)

Figure 1-1. *The relationship between AI, machine learning, and deep learning*

Machine Learning

Machine learning makes up a large portion of artificial intelligence being applied in businesses today. The goals of machine learning are to automate processes in order to decrease human effort, and to discover complex patterns that humans cannot interpret on their own.

This analogy is not perfect, but you can think of it this way: machine learning is to programming as the sewing machine is to sewing. Before the advent of the sewing machine, every stitch was sewn by hand. Once the sewing machine was introduced, sewing became faster, because not every stitch was handled by a human. With machine learning, we can build programs that handle far more complexity without having to hand **code** every detail. Ultimately, however, seams can't sew themselves, and machine learning continues to require a human hand to make it work.

In machine learning, machines are used to identify patterns in data and frequently predict the values of nonexistent data, often correlating to events happening in the future. Machine learning encompasses many methods for learning from data and makes up a large portion of research happening in artificial intelligence today.

What Is Intelligence?

The true sign of intelligence is not knowledge but imagination.

—Albert Einstein

While we intuitively know what intelligence is, it turns out to be difficult to summarize or formally define. There are many theories and definitions about what makes humans intelligent. How to measure intelligence has been argued by philosophers for centuries.

Shane Legg and Marcus Hutter collected over 70 experts' definitions of intelligence in a paper called "A Collection of Definitions of Intelligence." In an effort to derive a single definition, they came up with this: *Intelligence measures an agent's ability to achieve goals in a wide range of environments.*

In artificial intelligence, systems are often designed to mimic behaviors of the human mind. Researchers look to the human mind as a model of intelligence. The original goal of reconstructing human intelligence in machines requires teaching machines to carry out many complex functions. Reasoning, problem solving, memory recall, planning, learning, processing natural language, perception, manipulation, social intelligence, and creativity are all part of reaching this goal.

The Turing Test

How can we know if a machine is intelligent? The **Turing test** (**TT**) was proposed by Alan Turing in 1950 as one of the first tests of intelligence in machines. It is a challenge to understand whether a machine acts like a human. To pass the test, a human interrogator asks questions to the machine. If the human interrogator cannot distinguish which responses are from a human and which are from a machine, the machine passes the test.

The Turing test has appeared time and time again in popular science-fiction movies over the past 40 years. *Ex Machina* and *Blade Runner* are examples. It is one of many "Are we there yet?" checkpoints for the field.

How Machines Learn

Making mental connections is our most crucial learning tool, the essence of human intelligence; to forge links; to go beyond the given; to see patterns, relationships, context.

—Marilyn Ferguson, author

Understanding human behavior is complicated because humans do not always act rationally or logically. We can improve a machine's ability to predict human behavior by searching for patterns. These patterns help to discover and define trends. By analyzing these trends and modeling them with algorithms, machines can mimic human responses to certain inputs. Then, when encountering these inputs in real-world contexts, they are able to respond accordingly.

What Is Learning?

If we could simplify human learning, we might say that humans take information from their environment, relate it to something, and then learn or act. These inputs could be something they see, smell, taste, hear, feel, or even their interpretation of a mood or tone. That information is related to prior knowledge a person has about the world, making a connection. From there, a human might act on their new knowledge, explore, or innovate. This process can be observed in Figure 1-2.

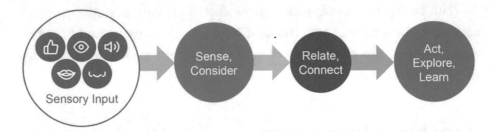

Figure 1-2. *How humans learn*

Machines are given input in the form of data. The machine interprets that data and learns from it. Then the machine evaluates that data before outputting the information that has been defined as useful to a human to interpret. This is the prediction phase, as shown in Figure 1-3.

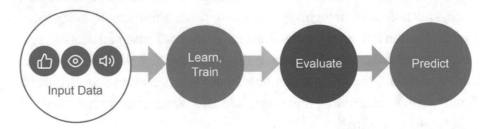

Figure 1-3. *How machines learn*

Where does the data come from? Machines are collecting data through **hardware** inputs as well as **software** programs. You can think of the hardware as the body, and software as the mind of a machine. Hardware addresses the area of **machine perception**, and software addresses the idea of both **machine language** and human language.

Machine Perception

Machines can perceive the environment through sight, feeling, and hearing via **sensors**. Sensors are a part of a machine's hardware system. They measure physical events like temperature, pressure, force, acceleration, sound, and light.

In fact, your phone can measure almost all these things. Phones sense through small electronics called **microelectromechanical systems** (**MEMS**). The microphone, camera, **inertial measurement units** (or **IMUs**, which help track position), and **proximity sensors** are all examples of MEMs. These sensors can also be found in various Internet of Things (IoT) devices.

In collaboration with these sensors, software systems on a machine can do things like interpret when a phone is upside down or right side up, measure human locomotion, and detect faces or sounds.

Language

Human languages are critical for communication. We use words and phrases, combining them in multiple ways in order to express ideas and emotions. Machines use machine languages to define models and parameters. Human language and machine perception both provide inputs in the form of data for machines to use to learn from.

An important distinction exists between machine languages and human languages. Machine languages are written in code. Originally, this code was only a series of 0s and 1s, or binary. Different combinations of 0s and 1s encode different information to machines. Over time, humans have created programming languages that interface between human language and machine language to make the job of coding easier.

The output of a machine is most useful when it can be interpreted by a human, which makes human language a useful concept for machines to understand.

Topics in Artificial Intelligence

The role of the computer is not to displace human creativity but rather to amplify it.

—Ray Kurzweil, *The Age of Intelligent Machines*

Successfully applying AI today requires understanding which techniques should be used for solving a given problem. There is currently no single algorithm that will provide value in every aspect of the fashion industry. The term *AI* as an overarching category can be confusing because it often leads people to believe that AI is a mysterious black box that can solve any problem. In reality, it is made up of several application areas, tools, and techniques. Understanding the broader categories and more specific subcategorization of the field gives a picture of how it all fits together.

Application areas discussed in this book include:

- Natural language processing (NLP)

- Computer vision (CV)

- Predictive analytics

- Robotics

Some commonly used tools and techniques include:

- Neural networks

- Generative adversarial networks (GANs)

- Data mining

Not every topic in the field of AI is covered in this book. Because the categories often overlap, this book is written with more information being introduced cumulatively as you read through. Later chapters may rely on explanation from earlier chapters.

Application Areas

The phrasing "application areas" refers to the specific areas where machine learning tools and techniques can be applied. Natural language processing, computer vision, predictive analytics, and robotics might use some of the same techniques like neural networks to solve different types of problems. Those application areas can be further extended to industry applications.

Natural Language Processing

Machine language and human language meet in **natural language processing (NLP)**. NLP is a way for computers to comprehend human languages. Every day, our interactions on the Web—things we post on social media, text messages we write, and so forth—contribute to an ever-expanding mass of data. Of this, it is estimated that 80% of the 2.5 quintillion bytes of data created every day is **unstructured data.** This is written in free form, unorganized and historically hard to parse. We can use NLP to understand the content and context of this unstructured data, unlocking a rich treasure trove of information about ourselves.

Natural language processing is applied in multiple product categories including conversational shopping, AI customer service chatbots, and virtual assistants and stylists.

Computer Vision

Computer vision (CV) is used to process and analyze images and videos. CV automates tasks we might associate with the human visual system and more. Although computer vision is a field of its own, artificial intelligence has played a major role in recent progress. Computer vision is frequently used in fashion applications because the fashion industry is so visual.

In the fashion industry, computer vision is being used in technologies such as visual search, smart mirrors, social shopping, trend forecasting, virtual reality, and augmented reality.

Predictive Analytics

AI can identify upcoming trends faster than industry insiders to enhance the design process.

—Avery Baker, chief brand officer, Tommy Hilfiger

11

Predictive analytics uses a variety of methods that use historical information to predict events that will happen in the future. These methods range in complexity and include data mining, basic statistics, and machine learning.

In this book, predictive analytics show up in two other areas: recommender systems and demand forecasting.

Recommender systems are part of predictive analytics. They seek to understand user or customer behavior and recommend products or services that the user is likely to like or purchase. Recommender systems have played a critical role for discovering products in e-commerce. You'll find them everywhere, from fashion retail web sites to behind the scenes in subscription box services. You'll also notice them in other areas including music and video streaming on sites like Netflix or YouTube.

Demand forecasting is used to optimize supply-chain planning. By predicting demand for products, the fashion industry can reduce overproduction, thereby cutting costs and reducing waste.

Robotics

Robotics, especially in apparel manufacturing, is a unique area of study that requires domain expertise across fashion, mechanical engineering, and machine learning. Robots have been used in industrial settings for many years in manufacturing for automotive, aerospace, and other industries that deal with mostly rigid parts.

Robotic manufacturing in the fashion industry is still a nascent field because of the complexities involved with handling fabrics. Nonetheless, with improvements in computer vision and in the planning algorithms needed to perform complex tasks, robotics is being adopted in fashion.

Tools and Techniques

Tools and techniques like neural networks, generative adversarial networks, and data mining are used across application areas. These methods are constantly changing and evolving to return higher quality results in industry.

Neural Networks

Neural networks are a subcategory of machine learning. They were originally modeled after our understanding of the behavior of neurons in the human brain: in the brain, a single neuron takes in input, processes it, and sends output. Neuroscience has moved away from this idea. We now know brains don't actually work like this, and the statistics behind neural networks in machine learning have been developed independently of neuroscience.

Neural networks are typically created with layers that compute information in parallel. They're composed of interconnected **nodes**. Knowledge in these systems is represented by the patterns that are taken on by nodes passing information to each other.

The way people think about the composition of a neural network usually includes three basic parts:

- *Input layers*: Contain input data

- *Hidden layers*: Contain the synapse architecture

- *Output layers*: Provide results from the network

Within that framework, a neural network can take on many architectures. Not all neural networks are the same. In the implementation, **training** is also an important part of the process. Training involves sending data through the neural network. In this stage, the network is learning complex connections between inputs and desired outputs. In many instances, the network's effectiveness is reliant on high-quality data.

Neural networks are frequently used in the application areas discussed earlier in this chapter. Understanding the basic mechanisms of neural networks helps provide a foundation for understanding how contemporary artificial intelligence works.

Generative Adversarial Networks

Unsupervised learning can be inefficient because machines must learn by themselves. What is obvious to us may not be obvious to a machine. **Generative adversarial networks (GANs)** are one way to increase the efficiency of unsupervised learning. GANs use two neural networks: one network generates results, and the other evaluates the accuracy of those results.

GANs are a more recently adopted technology in the machine learning space and have been proposed by companies such as Amazon as a method for creating AI fashion designers in 2017. These and other generative models are especially promising for creating unique new images as well as for filling in information from images that are incomplete or damaged.

Data Mining

Data is critical to any task in machine learning. Without data, the machine has nothing to train from. Data can include information such as video, images, and text. **Data collection** refers to the process of collecting data for analysis.

In many cases, data collection is just the beginning. What do you do with all the data? **Data mining** is about uncovering useful information in large amounts of data. For the fashion industry, social media can be a treasure trove for learning about the way customers feel about products and trends.

INTRODUCING BETTY & RUTH

Throughout the book, you'll find mention of a fictitious women's fashion brand called Betty & Ruth. Through the narrative of this brand, there will be examples of how certain tasks are currently handled at the fashion company and how they might be improved using techniques discussed in the book.

The examples from Betty & Ruth are an opportunity to explore implementation and sometimes expand on how other related technologies might fit into the picture.

Summary

Different application areas, techniques, and tools have strengths and weaknesses at specific tasks. Application areas of AI are often targeted at addressing specific needs (for example, image- vs. language-related problems are addressed using computer vision and natural language processing respectively). Successfully applying AI today requires understanding which techniques and tools make sense for your application.

Computer vision is an inherently visual field of artificial intelligence and is applied in instances dealing with images and video. Natural language processing, on the other hand, deals with communicating between human languages and machine languages.

Artificial intelligence, and especially machine learning methods, use data and models to understand and make predictions about questions we don't have the answers to.

For a deeper dive in the basics of artificial intelligence, I recommend the following books. You can find even more in the annotated bibliography and in the upcoming chapters of this book.

- *Artificial Intelligence: A Modern Approach,* by Stuart Russell and Peter Norvig (Pearson, 2016)

- *Artificial Intelligence: The Basics* by Kevin Warwick (Routledge, 2011)

Terminology from This Chapter

Artificial intelligence (AI)—A field of computer science that aims to teach machines to behave intelligently.

Code—Languages that are interpretable by machines.

Computer vision (CV)—A field of computer science dealing with the visual system. This includes teaching computers to process, analyze, and understand images and videos.

Data—Information that can be measured, collected, reported, and analyzed. It can take the form of various media, including text, images, and video.

Data collection—A process of gathering data for analysis.

Data mining—Uncovering useful information within a large dataset. *Refer to Chapter 9 for more information.*

Deep learning—Machine learning methods, usually in large neural networks that have more hidden layers, which increases the complexity of the relationship between input and output.

Demand forecasting—Encompasses multiple methods for predicting future demand for a product or service.

Generative adversarial networks (GANs)—A method of unsupervised learning using two neural networks in tandem to generate results and then analyze the accuracy of those results.

Hardware—The physical components of a computer system.

Inertial measurement units (IMUs)—Used to measure physical forces, angles, and sometimes magnetic fields around an object. In your phone, the screen rotates based on the phone's physical orientation, which is determined via IMUs that provide information about the phone's position in space.

Machine language—Language that is used to give machines specific instructions about what to do. In contrast to programming languages, humans typically can't read machine languages.

Machine learning—An application of artificial intelligence with the goal of modeling data patterns.

Machine perception—Refers to a machine's ability to take in information from its environment through the use of sensors.

Microelectromechanical systems (MEMS)—Really small electronics made up of components that are 1 to 100 micrometers in size.

Natural language processing (NLP)—Uses artificial intelligence to teach machines to use languages that are spoken and written by humans.

Neural networks—Also referred to as *artificial neural networks*, these are organized in a way that is similar to the way neurons work in the human brain.

Node—In a neural network, a node refers to the computer-based representation of a neuron. *Node* generally refers to a basic unit of a network in computer science. For example, your cell phone is a node in a network of cell phones.

Proximity sensors—Can determine whether something is nearby. Proximity sensors in your phone let your phone know whether your face is next to the screen while you're taking a phone call. This function means your phone doesn't turn off when your face touches it.

Recommender systems—Recommend products and services based on predictions about what a user will like or purchase.

Sensors—Used to give machines the ability to perceive the environment around them.

Software—The part of a computational system that uses machine languages to tell the machine what to do.

Training—The process in which a network or model is learning based on a particular dataset.

Turing test—A well-known test for determining whether a machine has human intelligence.

Unstructured data—Also referred to as *free-form data*, this is data that does not have a set structure (for example, a database) to help machines parse it.

PART II

Shopping and Product Discovery

CHAPTER 2

Natural Language Processing and Conversational Shopping

If you're not interested, you're not interesting.

—Iris Apfel, interior designer, fashion icon

Natural language processing (NLP) plays a critical role in human-machine communication. Billions of **gigabytes** of data are being created by users around the world every day. Most of this content is created in unstructured formats, making it unusable using regular programming techniques. With NLP, this unstructured data can be interpreted by machines without the requirements of strict data structures. To learn more about data and **data structures**, check out Chapter 6.

In the fashion industry, natural language processing has been used in applications like **conversational commerce**, chatbots, AI-based stylists, image and trend classification, and micro-moment shopping.

© Leanne Luce 2019
L. Luce, *Artificial Intelligence for Fashion*,
https://doi.org/10.1007/978-1-4842-3931-5_2

Finding garments you like on the Internet can be really hard, requiring sifting through tens, hundreds, even thousands of listings. One of the most pervasive concepts to gain traction in fashion retail is conversational commerce, which took hold in late 2017. By integrating product information into a chat interface, brands are able to reduce friction during product discovery and provide highly personalized experiences to consumers searching for products, information, and customer service.

This chapter gives context for the emergence of conversational commerce and natural language processing, the technology that enables it.

Natural Language Processing

Natural language processing has been studied by computer scientists since the 1950s. Computer scientist Alan Turing thought the ability to use human language to be an important determinant of intelligence in machines. He later created the Turing test as a measure of machine intelligence. As noted in Chapter 1, a machine passes the Turing test if it can fool people into believing it is a human. At that time and throughout the 1960s, the first **chatterbots** were created, exemplifying the power of natural language–based interfaces.

ELIZA

What makes you think I am entitled to my own opinion?

—ELIZA, chatbot

One of the most famous examples during this time was **ELIZA**. ELIZA was one of the first programs to pass a restricted version of the Turing test. Simulating a **Rogerian psychotherapist**, the program processed user inputs, saving them in memory and recalling them during conversation.

ELIZA was originally created by Joseph Weizenbaum at the MIT Artificial Intelligence Laboratory.

The ELIZA bot consists of a long list of possible responses and complex rules to determine which responses are used in conversation. In the 1980s, the architecture of these bots all changed because of machine learning. **Chatbots** today are capable of more complex interactions because of the algorithms that control them.

Note   To interact with a sample ELIZA, visit `https://thefashion robot.com/eliza`.

Chatbots

Most chatbots can be placed into two basic categories: scripted and artificially intelligent. **Scripted chatbots** can follow only a predefined set of rules. This set of rules means the kinds of questions the chatbot can answer and the responses it can create are limited to the scripts it was programmed with. Artificially intelligent chatbots are built to interpret natural language used by humans and are capable of coming up with relevant responses to inputs that are not exactly pre-defined.

More recently, chatbots can use images as part of a conversation in addition to text. The computer vision techniques that enable visual search and other image-based features are discussed in Chapter 3.

Specialized Chatbots

Although numerous companies create general-use chatbots, some companies create chatbots specifically for retail applications. These chatbot services are more likely to help fashion retailers because a general-purpose chatbot may get confused when discussing fashion or retail with a customer.

Companies like mode.ai bake in **style preference** as well as **size and fit preference**. Another feature some of these companies are working on is integrating cross-brand size correlations to make it easier for consumers to know what size to buy in the moment that they're considering purchasing.

In the end, it seems like these specialized chatbot services will become a one-stop shop for fashion brands looking for AI-assisted product discovery, product care, and customer service.

Conversational Commerce

> *I don't know anyone who likes calling a business. And no one wants to have to install a new app for every business or service that they interact with. We think you should be able to message a business, in the same way you would message a friend.*

> —Mark Zuckerberg at F8 in 2016

Conversational interfaces aren't new. Although they have been around since the introduction of chatterbots like ELIZA in the 1960s, their traction today is likely explained by the rise in popularity of messaging apps. According to Business Insider, in 2015, messaging apps outpaced social media apps in growth. Messaging apps, which did not have widespread adoption until recent years, make for a natural interface for chatbot conversations.

Natural Language Queries

The main idea behind conversational commerce is to reduce the number of clicks that a user has to go through to reach a desired product. Rather than selecting a half dozen filters, as in Figure 2-1, a user can type what they're looking for in a **natural language query**. Figure 2-1, which requires at least four clicks to get to the product query, could be replaced with the simple input *Find Women's Sandals with 3-4" Heels in Black under $100.*

Price	Styles	Color	Heel Height
☐ $50 and Under	☐ Sandals	☐ Black	☐ Flat
☐ $100 and Under	☐ Boots	☐ Brown	☐ Under 1in
☐ $200 and Under	☐ Sneakers	☐ Blue	☐ 1in - 1 3/4in
☐ $200 and Over	☐ High Heels	☐ White	☐ 2in - 2 3/4in

Figure 2-1. Search filters that you might find on a fashion e-commerce web site

Shopping and Messaging

Conversational shopping, or conversational commerce, seems to have emerged as a concept in 2015. The AI bots behind conversational commerce interfaces act as **agents** on the other side of a messenger conversation with a consumer. By asking questions to the bot, a consumer can receive personalized recommendations, product care instructions, customer service, and even purchase products in one click.

The term *conversational commerce* has been attributed to technologist Chris Messina who described it as follows:

> *Conversational Commerce pertains to utilizing chat, messaging, or other natural language interfaces (i.e., voice) to interact with people, brands, or services and bots that heretofore have had no real place in the bidirectional, asynchronous messaging context.*

> —Chris Messina, founder, Molly

A conversational interface does not necessarily include only messaging. It also encompasses buttons, web views, images, and other simplified **graphical user interface (GUI)** components. These components can help guide the conversation between the human user and the machine by providing possible outcomes to the specific context.

To mimic the experience of talking with an in-store sales associate, companies like Levi Strauss & Co. have partnered with AI companies like mode.ai. In late 2017, Levi's and mode.ai released a conversational commerce bot that helps consumers discover their perfect jeans. Figure 2-2 shows a screenshot of the conversational shopping interface by mode.ai that appears on Levi's Facebook Messenger and web site. This particular example uses a mixed UI relying not only on the user's improvised inputs, but allowing the user to select common options by clicking preexisting buttons.

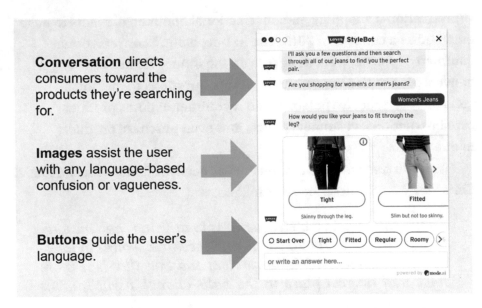

Figure 2-2. *A conversation between the Levi's Messenger bot by mode.ai and a consumer searching for jeans*

Personalized Shopping Experiences

Part of what makes conversational commerce so enticing is the ability to create customized experiences for users. Particularly for younger generations, personalized experiences are predicted to dramatically shift online purchasing. According to McKinsey, "Personalization can deliver five to eight times the ROI on marketing spend, and can lift sales by 10% or more."

AI agents learn as a consumer interacts with them, making suggestions for products and actions based on products they've liked. At times, these agents even adapt the GUI based on click-rate. For example, UI components like buttons may be revealed, like the ones shown in Figure 2-2. As users interact with the bot, data is collected about engagement with each UI component. These components can be adjusted in real time, depending on their success with customers.

Bot-to-Bot Interaction

For brands, the proposition of hosting bots that interact with other bots makes revenue sharing easier. As Karen Ouk from mode.ai has pointed out, a T-shirt brand bot might recommend pants from a pants brand bot, and this cross-recommendation can help both brands expand their reach and grow.

The idea of bots having business relationships with one another might sound absurd now. As these bots begin generating real revenue for each other, it may be difficult to imagine a time when brand partnerships were marked by sweepstakes and pop-up shops.

Context-Based Decision Making

An AI-based agent has the opportunity to not only provide product-based recommendations but also tailor recommendations to the user's context. Examples of the user's context include location, language, and demographic data. An agent should make different recommendations for winter clothes in New York than for winter clothes in San Francisco.

In the future, the context for these AI-based conversations will become more correlated with the user. It could sync with user calendars, suggesting outfits for the holiday party they're going to attend tomorrow or the board meeting they have on Friday, providing a full-blown comprehensive fashion assistant. Refer to Chapter 5 for more details on the fashion and style assistant concept.

Live Chat

In 2017, the National Retail Federation's Omnichannel Retail Index called **live chat** one of the fastest growing areas of omnichannel retail. According to the report, 54% of retailers have implemented live chat on their web sites. Many businesses incorporate live chat on their web sites in order to answer in-the-moment customer service questions, send order information, and more.

For consumers, talking to a sales associate in a shop is a way of getting a second opinion. The ability to interact with customers while they're shopping online has been historically limited. The moments in which the consumer is making a purchase decision are critical to conversion rates.

Live chat, or sales chat, is just one way that retailers are using conversational interfaces to drive consumer decision-making. Live chat interfaces connect consumers with a human, an AI bot, or a hybrid solution. Historically, live chat meant talking to a human through a chat interface embedded on a brand's web site. By using AI bots, these one-to-one services are much more scalable. The distinction between live chat and conversational commerce bots will likely disappear as the implementations of these services become more streamlined.

BETTY & RUTH CHATBOTS

If you're a smaller brand like Betty & Ruth, you might think that AI bots are out of reach. It's normal in the fashion industry that we don't have things like an **application programming interface (API)** that other software companies can use to access our product data. (What APIs are and how they work is explained in Chapter 9.)

Fortunately, the companies working on conversational commerce have recognized that fashion brands don't always have tech resources. As a solution, for brands that agree to participate, they've built **site crawlers** that will pull product information from the brand web site for use in their AI-bots. Preparing a web site for basic integration doesn't require brands to implement any new technology.

The downside is that most of the out-of-the-box solutions can't address all of the needs we have as a brand at Betty & Ruth. With the current state of things, using third-party apps, we would need more than one specialized chatbot to address product discovery and customer support.

There are other bots we are looking into integrating that are more general-purpose bots and focusing on chat features like cart recovery, which addresses a pain point we have: abandoned carts.

How Machines Read

Natural language processing plays a critical role in many applications. Imagine doing **sentiment analysis** to learn what your customers are saying on social media about ripped jeans this season.

But what is the computer doing when it dissects words? It usually starts with normalizing the text, which is a process that transforms natural text into data the machine can understand. For example, it could start with correcting misspelled words. Then it follows through a process of breaking the words apart, analyzing them, and relating the words to each other in order to extract meaning. Four commonly addressed methods happening behind the scenes in NLP are as follows:

- Tokenization

- Word embeddings

- Part-of-speech tagging

- Named entity recognition

Tokenization

In the process of **tokenization**, a machine breaks apart pieces of a sentence or phrase into tokens, usually words or terms. This process is also sometimes referred to as **lexical analysis**, or **lexing**. These linguistic units make up words, punctuation, numbers, and so forth.

Tokenization is an important preprocessing step preceding categorization of words. Figure 2-3 shows an example of how a system might separate tokens by the whitespace, or spaces, between words.

This is a sentence.

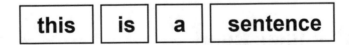

Figure 2-3. *A machine might tokenize "This is a sentence" by using whitespace as a separator*

Whitespace tokenization is just one of many ways to tokenize text. Other methods are used for more complex sentence structure or languages that don't use whitespace to separate words.

Word Embeddings

How can a machine understand or compare words? One way is to associate each word with a set of number values, or vectors, so that a machine can compare them. These vectors are known as **word embeddings,** or word vectors.

Word embeddings make words accessible to certain types of machine learning models, by making it easy to do mathematical calculations with them. They can help define relationships between words and create visualizations of these relationships.

Skip Grams

One way of analyzing these relationships and their proximity is by using **skip grams**. The skip gram model allows us to analyze how likely two words are to be co-occurring near each other in a text. The example in Figure 2-4 shows the sentence "Well-made dress, runs small." The input word is highlighted in pink. Looking at the nearest two words, the skip gram word pairs from this sentence are shown on the right. What this means is that for this item, the word pair "dress, small" is more probable than "dress, large."

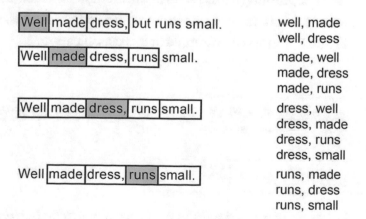

Figure 2-4. *Word pairs from a skip gram training model*

Part-of-Speech Tagging

Part-of-speech (**POS**) **tagging** is the process of defining the part of speech of a given word based not only on the word's definition, but also its context. The English language contains eight parts of speech: nouns, pronouns, adjectives, verbs, adverbs, prepositions, conjunctions, and interjections. Knowing the part of speech of a word reveals a lot of information about its neighbors and helps to comprehend the sentence as a whole.

Often, tokenization is done as a precursor to this task in order to separate the words to be tagged. The process of tagging is a disambiguation task, making the ambiguity of words clearer within their context. Figure 2-5 shows an example of part-of-speech tagging.

Figure 2-5. *Part-of-speech tags from the sentence "This is a sentence," which is tokenized in Figure 2-3*

Named Entity Recognition

Named entity recognition (**NER**) refers to the methods for identifying and classifying important nouns into categories such as organizations, people, and times. Figure 2-6 shows an example of this classification.

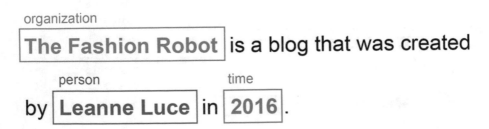

Figure 2-6. *Labeling named entities in the sentence "The Fashion Robot is a blog that was created by Leanne Luce in 2016" to classify nouns*

Natural Language Understanding

As a subtopic of natural language processing (NLP), **natural language understanding (NLU)** addresses challenges in comprehension of human languages. While NLP is about reading language, NLU is about understanding language. Aspects of understanding language such as sentiment analysis and relation extraction both fall under the subtopic of NLU. Figure 2-7 shows the relationship between NLP and NLU.

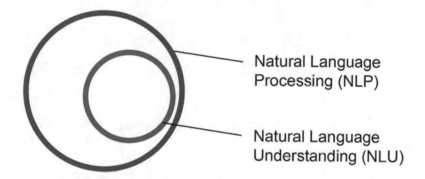

Figure 2-7. *Natural language understanding is a subtopic of natural language processing*

Sentiment Analysis

Sentiment analysis is a method of understanding how the speaker feels about a particular object or subject. There are a number of methods for understanding sentiment, including using machine learning, statistics, knowledge-based methods, or a hybrid of these.

While knowledge-based methods rely more heavily on determining sentiment in definite terms, machine learning methods allow for more flexibility in meaning. In the paper, "Recursive Deep Models for Semantic Compositionality Over a Sentiment Treebank," Stanford University researchers Richard Socher et al. used a deep learning model not only to analyze the sentiment of individual words, but to build a representation of the entire sentence structure in order to determine its sentiment. An example of the treebank is shown in Figure 2-8.

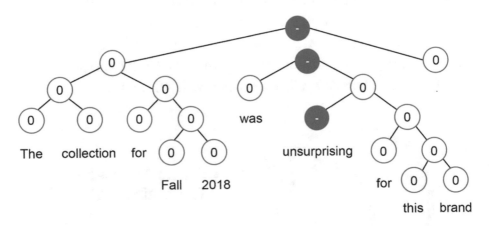

Figure 2-8. *Results of a treebank in a live demo created by Stanford researchers*

Based on this model, the sentence "The collection for Fall 2018 was unsurprising for this brand" reflects a moderately negative sentiment. Negative sentiment is attributed to the term "unsurprising." This is represented in the image by circles which contain a "-."

Relation Extraction

Relation extraction is a way of extracting specific pieces of information and their relationship to the subject from a text. There is a long list of methods used to achieve relation extraction, but the goals are approximately the same for all methods. Relation extraction looks to comprehend things such as the subject, relation, and object of a given sentence. Less specifically, it is looking for relationships in unstructured text. This is useful because as a machine reads through text, it can learn from the sentences in that text, store that knowledge to a database, and recall that information when prompted.

An example of how a machine might use relation extraction can be seen in the following text and Table 2-1.

*PVH is a corporation headquartered in **New York, NY**. It was incorporated on **April 8, 1976** as an **apparel company**.*

Table 2-1. *Extracting Relations from Natural Language into Structured Data for Later Use*

Subject	Relation	Object
PVH	Location	New York, NY
PVH	Incorporated	April 8, 1976
PVH	Is A(n)	Apparel Company

Summary

Natural language is the way that humans interface with each other to communicate ideas from coordinating tasks to shopping for our next fashion fix. Natural language processing lets machines understand and communicate using human language. In this new world of artificial intelligence, language as an interface makes finding that perfect pair of black skinny jeans faster and easier.

Chatbots may have started with simple scripts, but today they are capable of interpreting full sentences and images on the fly and returning relevant results immediately. Though they weren't always popular, it seems chatbots are here to stay. They increase engagement online, provide personalized experiences for discerning shoppers, and much more.

Thank you to Karen Ouk, chief business officer at mode.ai for answering questions about conversational commerce for this chapter.

Terminology from This Chapter

Agents—Also referred to as an *intelligent agent* (*IA*), this is an autonomous entity that responds to variable environments and works toward achieving goals.

Application programming interface (**API**)—Makes it possible to share data from a web site without revealing all the code or requiring developers to take action in sharing that data. They usually provide clean, easily referenced data that can be used on another site or app. For example, you might use an API to share your Twitter feed on your web site.

Chatbots—Machine-based entities that receive and respond to text- or auditory-based inquiries.

Chatterbots—Another name for a chatbot, chatterbot is no longer popular terminology. Some might suggest that chatterbots refer to a particular class of chatbots created before 1980.

Conversational commerce—Utilizes chat to connect consumers and products.

Conversational interfaces—Encompass chatbots from those that are purely conversational to ones that integrate buttons and other GUI features.

Data structures—Methods for organizing data in a machine so that it can easily be accessed.

ELIZA—One of the first chatterbots to pass the Turing test in the 1960s. It was named after Eliza Doolittle of *Pygmalion*.

Gigabytes (**GB**)—Units of storage on machines; it is a multiple of the unit byte. 1GB = 1,000,000,000 bytes.

Graphical user interface (**GUI**)—A GUI is a graphical representation of computer software which allows a user to easily interface with a computer.

Lexical analysis—The process of converting a series of characters into tokens. See also *lexing* and *tokenization*.

Lexing—The process of converting a series of characters into tokens. See also *lexical analysis* and *tokenization.*

Live chat—A tool that enables a business to engage with consumers in real time. Some live chat systems interface with a human, while others interface with intelligent agents, and some a hybrid of both.

Named entity recognition (**NER**)—The process of finding and labeling proper nouns and other names in a sequence of words.

Natural language query—A search query that is input in natural language. These interfaces are able to take in phrases of human language and interpret the nature of the query, and then return relevant results.

Natural language understanding (**NLU**)—is a subtopic of natural language processing specific to the understanding of human language.

Part-of-speech (**POS**) **tagging**—The process of defining the part of speech of a given word based not only on the word's definition, but also its context.

Rogerian psychotherapy—A type of person-centered therapy created by psychologist Carl Roger in the 1940s, practiced into the 1980s. The popular culture impression of this type of therapist can be captured in the phrase "... and how does that make you feel?"

Scripted chatbots—Made up of sequences of pre-determined scripts. They can respond with specific answers to specific questions, but have a difficult time breaking out of those specialized dialogues.

Sentiment analysis—Useful for determining the emotional impact of a set of words.

Site crawlers—A kind of software that scans a web site for information, reads the content, and turns it into structured data. They are frequently used for search engines.

Size and fit preferences—The size that a person prefers and the way they like a garment to fit. This could be characterized, for example, as "loose fit" vs. "snug fit."

Skip grams—Word pairs that are considered in the context of one another. There may be gaps between them made up of words that are skipped over during analysis.

Style preferences—A consumer's preference when choosing the style of a garment. In jeans, for example, this could refer to "skinny jeans" vs. "bootcut jeans."

Tokenization—The process of converting a series of characters into tokens (often words). See also *lexical analysis* and *lexing*.

Word embeddings—Real numbers or vectors that have been mapped to words. *Also referred to as word vectors.*

CHAPTER 3

Computer Vision and Smart Mirrors

To me, fashion is like a mirror. It's a reflection of the times. And if it doesn't reflect the times, it's not fashion. Because people aren't gonna be wearing it.

—Anna Sui, fashion designer

Smart mirror technology is sweeping through retail environments, from luxury department stores to personal living spaces. A smart mirror is a two-way mirror with an electronic display behind it. They are computers enabled by a full stack of technology, from hardware with depth sensing to software equipped with advanced computer vision algorithms.

These mirrors aren't just a fantasy. In fact, they often aren't even mirrors. The technology behind smart mirrors integrates cameras and two-way glass with a digital screen that provides distortion correction, object detection and recognition, feature extraction, and augmented reality. The magic of these computer vision techniques is that users don't realize how much is going on behind the scenes.

Computer vision, as mentioned in Chapter 1, refers to the vision system in a machine. Giving machines the ability to "see" relies on generating digital images or video by means of a digital camera for the machine to interpret. The computer vision concepts discussed in this

© Leanne Luce 2019
L. Luce, *Artificial Intelligence for Fashion*,
https://doi.org/10.1007/978-1-4842-3931-5_3

chapter transfer across many new technologies in the fashion industry. These concepts apply to applications involving images, video, and even three-dimensional digital assets.

Retail Meltdown

Before diving into the details and technology features of smart mirrors, I want to address the reason that retailers are looking to new solutions for their brick-and-mortar shops. In 2017 alone, there were 6,985 store closures, and more than 15 major retailers went bankrupt. With these closures came the consolidation of businesses through mergers and acquisitions. But today, fashion businesses are competing on more than just the race to the bottom.

Younger generations are demanding more-engaging experiences in everything they do. While the fashion industry has left no rock unturned in the physical world, these younger consumers are demanding to be delighted through digital. The future of retail relies on innovations that bridge the digital world, where they spend so much of their day, and the physical world, where clothing and other physical products reside.

Smart Mirrors

In difficult times, fashion is always outrageous.

—Elsa Schiaparelli, fashion designer

Retailers using smart mirrors are not only looking at implementing mirror hardware. They are looking at how adding this technology can bolster their omnichannel strategy by integrating in-store experiences with mobile devices. The smart mirror in these stores plays a potentially essential role in getting information to consumers about the products they tried on and how to buy them later. Figure 3-1 shows an image of a smart mirror in use at a luxury retail store.

Figure 3-1. *A woman trying on a dress using MemoMi's smart mirror with augmented reality features to change the color of the garment in real time*

Smart mirrors aren't just about showing users what they would look like in a garment across colorways, but also about showing the user in many types of garments. Smart mirrors transform the try-on experience from an annoying hassle to a delight.

BETTY & RUTH OMNICHANNEL STRATEGY

For Betty & Ruth, virtual try-on is not the most valuable feature that smart mirrors have to offer. These mirrors connect to our digital sales channels, e-mail marketing, social media, and other digital marketing and sales strategies by collecting data about each unique user. These mirrors are helping us to achieve our goal of having a seamless omnichannel experience that improves our customer experience and our brand consistency.

To make the experience of brick-and-mortar product discovery more effective, our customers can return home to their own computers with images and information about the products they tried in their smart mirror experience at our store. From there, they can make purchases at their own speed.

Data Collection

How long do customers spend looking at something? What colors did they try on? What colors did they ultimately buy?

Like any digital device, smart mirrors have the potential to give retailers more insights about what their users are looking at, looking for, and buying. With this data, retailers can learn how to better optimize for a personalized product discovery flow, deliver recommendations with higher accuracy, and ultimately increase conversion rates. Chapter 6 explores big data and what machine learning can do with large amounts of data.

Social Sharing and Checkout

Smart mirrors provide more than just a fun interface to try on clothes and more than just a data collection point for retailers. They also provide a quick-and-easy method for sharing images and videos directly from the retail location. In a retail environment without mechanisms for real-time sharing, customers are far less likely to post products they see in stores. Here, they can share items instantly and seamlessly, tagged with product information needed to make a purchase. Making it easy to share creates a better experience for the customer trying to find an item again, the customer's social network trying to buy the item, and the retailer who is more likely to convert that retail session into a sale.

Companies like MemoMi make it simple for consumers to send themselves videos, side-by-side comparisons of different outfits and colors, and more. Some of the greatest value adds in this instance might have less to do with virtual try-on and more to do with capturing real-world information through a digital device quickly and easily. An example of a user sharing a virtual try-on session can be seen in Figure 3-2.

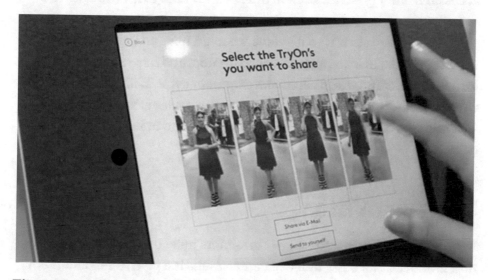

Figure 3-2. *Users can select images to share with themselves and friends instantly from an in-store smart mirror interface*

Smart mirrors can make the experience of shopping much easier by combining try-on with checkout. Customers can make purchases using smart mirrors that will arrive at their door later or leave the store with those newly purchased items. This means they don't have to wait in long lines or walk around the shop looking for the cash register.

Implementation

Implementation can be challenging with these devices and depends on the features that a retailer is trying to include. A full virtual try-on experience requires 3D digital assets for every garment. This currently can pose a challenge because most fashion design is not done in digital 3D. For this reason, features like changing the color of garments can be a more accessible entry point for fashion brands. This kind of feature does not require 3D assets.

If collecting data is a goal, retailers should have an understanding of what they'd like to learn about their customers and have mechanisms for feeding that into the product development cycle. Without this, it is worthless information to collect. Chapter 6 includes examples of how subscription-based businesses rely on user data to inform their product decisions.

Computer Vision

In order to understand an image, computers must translate an image into numbers. The basic unit of a digital image is a pixel, a set of numeric values that correspond to a color. For example, in an 8-bit black-and-white image, the number 0 would display as black, 255 as white, and a value in between like 200 as a light gray. Machines use the value from that pixel and surrounding pixels to apply algorithms to the image. Mathematical calculations based on patterns in these values are what machines use to determine the content of an image.

Figure 3-3 shows what a 10 × 10 pixel image might look like to a machine. Each pixel has a numeric value ascribed to it, visible in the lower right.

Figure 3-3. *Left, image by Michael Rubin. Top right, enlarged piece of the image. Bottom right, values per pixel.*

Image processing refers to a broad set of techniques used to manipulate an image. These techniques are used in a wide variety of applications by photographers and engineers alike.

The main difference between standard image processing and image processing in computer vision lays primarily in the goal. In computer vision applications, the goal is generally to enhance an image to make it more readable by a machine.

Transformation

By using the numeric values that represent a digital image, machines are able to apply mathematics to images in **image transformations**. An example image transformation, shown in Figure 3-4, adds the value of 20 to each pixel on the grid. In grayscale images, lower pixel values closer to 0 are darker, and higher values closer to 255 are lighter. The result is that the entire image is lightened.

Figure 3-4. *An image transformation that adds +20 to each pixel. Transformations can also refer to cropping, rotating, and applying other simple filters to make an image easier for a machine to read.*

Filtering

Filtering is usually used to enhance an image, extract information, or detect patterns before sending it through a machine learning model. By applying filters to images, machines may have an easier time with more-complicated tasks down the road.

Filters are algorithms that can be applied manually through software like **Adobe Photoshop** or programmatically using a variety of programming languages or specialized applications like **Matlab**. The advantages of using code rather than a manual process are that the filter can be reused and activated automatically.

There is more crossover between applications like Photoshop and Matlab than you might think. Many professionals who are trained in fashion design or fashion photography are familiar with Photoshop filters.

In Photoshop, a **Gaussian blur**, seen in Figure 3-5, is the same as it is in computer vision. A Gaussian blur is a type of low-pass filter frequently used to reduce noise in images. It works by calculating the average between a pixel and a cluster of its neighbors, giving the blurred effect.

Figure 3-5. *The effects of a Gaussian blur, filter applied to a photograph using Adobe Photoshop*

Imagine you have a folder of 1,000 images. Each of those images needs to be rotated and cropped to the same dimensions, resized to the same pixel ratio, and filtered to black-and-white. You can do this manually, but it will take a long time, probably hours. You can also do this programmatically, and it will take minutes, even seconds. In machine learning, this kind of example is just a preprocessing step to prepare the images before a machine learning model can be applied.

Feature Extraction

In computer vision, a feature is vaguely described as an interesting part of an image and more precisely as a higher-level description of raw data. **Feature extraction** refers to processes used to find features. Features include edges, corners, interest points, blobs, and ridges. Finding features allows a machine to also define a local segment in an image where an object or area of interest might be found.

All feature detection is considered to be low-level processing, which means that it happens early on in the process of applying computer vision algorithms to an image. Features aim to isolate interesting parts of an image, though not all features achieve that. Consequently, it's one of the most important steps to get right. Without a repeatable and reliable feature-detection process as a base, all the following algorithms applied to that image may be less accurate.

Edge Detection

Edge detection is a technique that finds the boundaries of an object within an image. Edge detection, an example of which is shown in Figure 3-6, is a foundational tool used in feature detection, feature extraction, establishing key points, removing backgrounds, and more.

Figure 3-6. *An image of a shoe on the left and an edge-detection algorithm applied to the same image on the right*

Object Detection

Object detection is used to determine whether objects are present in an image. An object could be a pencil, a shoe, an elephant, and many things in between. At the object-detection stage, however, the computer does not know specifically what the object is.

Once an object is detected, **localization** can be used to narrow down where the bounding box of an object is within an image. When the image is used in a neural network or other machine learning models, localization can help reduce the computational load by reducing the part of the image that needs to be interpreted to a smaller area. After finding the object and its bounding box, the object can be classified. Figure 3-7 shows localization of an object within an image.

Figure 3-7. *Localization of an object—in this case, a shoe—within this image*

Image Classification

Image classification is one of the most explored and well-known problems in computer vision. It refers to the process of classifying images into one of many predetermined possible categories. One of the major limitations of image classification is its reliance on a dataset of labeled images for training. Creating this data usually requires an extensive amount of manual labor.

Neural networks are often used in image classification. We'll return to neural networks and how they solve problems like this in Chapter 4.

Beyond Static and 2D Images

Beyond static images, computer vision can be applied in order to understand video and 3D objects. In video, motion can be used to track walking patterns of a crowd. This kind of analysis reveals areas of interest on a retail floor, where customers stop and look more often.

Smart mirrors and similarv camera devices, especially those equipped with depth sensing, are able to capture 3D information. Down the line, these methods could be used to provide more adequate fit-matching or garment size-customization to consumers.

Summary

Smart mirrors are unlocking potential in omnichannel retail, data collection, in-store engagement, social sharing, and augmented reality. They provide an experience that feels custom to the user and provides retailers with more insights about what their customers are looking for.

Using computer vision and machine learning methods, these mirrors are providing a new bridge between brick-and-mortar retail and the rapidly expanding e-commerce marketplace. Knowing what is happening behind the scenes in these devices can reveal where new insights can be captured and spark new ideas about what is possible!

Terminology from This Chapter

Adobe Photoshop— A popular and widely used software program in the creative industry for manipulating digital images.

Edge detection—An image-processing method for finding the edges of objects in digital images.

Feature extraction—The process of identifying and quantizing a feature, often with the goal of extracting higher-level information.

Filtering—Used to enhance an image, extract information, or detect patterns. It is a commonly done before sending images through a machine learning model.

Gaussian blur—A commonly applied low-pass filter that is used in image processing, from photography to machine learning.

Image classification—The process of classifying images into one of many predetermined possible categories. Depending on the application, these categories can be anything from animal species to types of lace.

Image processing—Using computer algorithms to manipulate images in order to enhance them or extract more useful information from them.

Image transformation—Can refer to mathematical calculations that result in translation, rotation, scaling, cropping, and shearing of an image. In computer vision, this is usually used to correct the alignment or distortion of an image.

Localization—A process for finding the most important part of an image. By reducing an image down to areas of interest, localization helps to reduce the computational load.

Matlab—A commercial software package often used by computer programmers to develop computer vision algorithms.

Object detection—The ability to detect objects such as humans, shoes, or handbags in digital images.

Smart Mirror—A computer equipt with a camera and 2-way mirror in front of it. These devices are often used in brick-and-mortar retail locations to recommend garments and allow users to virtually try-on garments. They are also sometimes used in the home.

CHAPTER 4

Neural Networks and Image Search

I envision some years from now that the majority of search queries will be answered without you actually asking. It'll just know this is something that you're going to want to see.

—Ray Kurzweil, author, director of engineering, Google

It's hard to imagine an industry that relies on images more than the fashion industry. Almost every process, from manufacturing to marketing, revolves around images. This chapter discusses methods for classifying images, developments in neural networks that have been improving these methods, and the basics of how neural networks work. The idea of classifying images was mentioned in Chapter 3. It might not sound like a futuristic or exciting concept, but it is foundational for machines to answer the question, "What is this?" when working with an image.

When might you want to know what is in an image? In retail, the ability for customers to search for a particular garment style on a web site gives them access to the products they are searching for. Even better, the ability to discover products from styled images gives customers the ability to browse inspiration (what to do with a product or how to wear it) and to access the product (which item to buy to achieve the desired look).

© Leanne Luce 2019
L. Luce, *Artificial Intelligence for Fashion*,
https://doi.org/10.1007/978-1-4842-3931-5_4

Fashion Industry Images

From marketing materials to design tools, images have a huge impact on operations in the fashion industry. Images are highly important in designing, constructing, and selling fashion. The ways images are used in the fashion industry include the following:

- Fashion photography
  - High-fashion images
  - Lookbooks
  - Editorial
- Web imagery
  - Product photography
  - Social media images
- Design drawings
  - Technical flats
  - Tech packs
  - Material references
  - Construction detail references

As you're reading this chapter, can you imagine how having a machine with the ability to identify the content in these images could be helpful?

Image Search

The possibilities for using image search extend to the many aspects of the fashion industry that rely on images for information. Figure 4-1 shows three types of image-search techniques: image search, reverse image search, and visual search. With given text, a **search engine** is able return images that have been tagged with matching and related **keywords**.

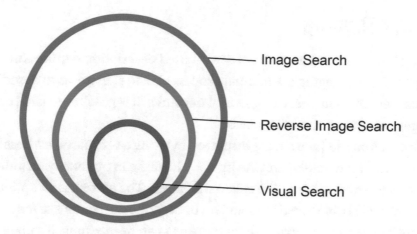

Image Search

Reverse Image Search

Visual Search

Figure 4-1. *A chart of three types of image-searching techniques and their relationship to one another*

Image search refers to the general topic of finding images, but usually refers to a search process based on a text input. The concept of image retrieval was introduced in the 90s, but wasn't popularized until image search was introduced by Google in 2001, after Jennifer Lopez's green Versace dress sent Internet users into an image search frenzy.

Reverse image search is a subset of image search referring to a search query in which an image is used to find another image. A further subset, **visual search**, refers to a process of finding items within an image and searching for those. For example, when searching for an image of a fashion blogger wearing a pair of black pumps, the search results will return the black pumps rather than returning more images that are visually similar to the image of the fashion blogger.

Image search is not a cutting-edge idea and doesn't necessarily include AI. However, in the fashion industry, even basic image search is hardly viable as a search tool. The way that images, from inspiration to process drawings, are organized is more often through folders on desktops rather than in any structured company-wide database.

Image Tagging

To be able to search images from a text-based description requires **image tagging**. Image tagging is a manual process of using keywords to describe the content of an image. Using neural networks, it is possible to label or tag images with fewer manual processes.

Images on the Internet are displayed by using a text-based address to tell a computer where to find the image. The text can optionally include other descriptive information called **metadata**. The **alt text** is an optional element that is often used in metadata to describe the content of the image. Not all images include it, and the text can be anything. It's up to the programmer to make it a description of the image. The **HTML code** in Listing 4-1 shows the basic **markup**.

Listing 4-1. HTML Code That Machines Use to Render an Image on the Web

```
<img src="/location/newyorkstatueofliberty.jpg" alt="New York
Statue of Liberty" />
```

In the case of this HTML code, the machine finds the location of the image listed after src, short for *source*. Keywords listed in the alt text help machines recognize the image content. In this example, search engines could identify that the image is of the Statue of Liberty, located in New York City, based on the alt text that provides that information. Metadata can be used as the thing being searched, or images could be used as the thing being searched. The way the search process is undertaken may or may not use machine learning.

Reverse Image Search

Reverse image search is an image-search method popularized by Google in 2011. In this method, an image is used as the search input. It is then analyzed; a search query is created, and results are given to the user.

The query is generated by a combination of factors including the image file name, link text, and text near the image. Figure 4-2 shows results from Google's reverse image search.

Figure 4-2. *Reverse image search: an input image and visually similar results from Google*

This search method might be used to track down the source of an image, find web sites that an image is posted on, get information about the image, find higher-resolution versions, or access images with similar content. Before the introduction of this image-search method, screenshots from the Web would remain un-citable artifacts on user desktops.

Reverse image search also used computer vision algorithms for object recognition and to extract other visual information. Computer vision methods are discussed further in Chapter 3.

Visual Search

Visual search, also implemented through the use of computer vision, provides us with the ability to search through troves of visual data without relying on text. Similar in concept to the challenges faced in natural language processing, the ability to search images that have not been tagged unlocks access to images that would otherwise not be found. Visual search takes an image as an input and returns similar images based on visual characteristics in the image. Figure 4-3 shows an example of visual search results.

Figure 4-3. *Search results on Neiman Marcus using visual search by Slyce*

While reverse image search is optimized for similar images, visual search can be optimized to search for similar items across images. There are a range of ways to create a visual search model.

Neural networks (NNs) are a mathematical or computational tool. Computer vision is a field that applies that tool to image data. The computer vision system is able to recognize objects in the image. The ability to identify that the person is walking and what kind of heels she is wearing is likely a task for machine learning. The key difference here is that computer vision gives the ability to see, while machine learning (namely, neural networks) gives the ability to recognize objects. Computer vision and machine learning can be used to solve problems independently, but more can be accomplished when the two are combined.

BETTY & RUTH IMAGE SEARCH WORKFLOW

Image search can be useful in unexpected ways. At Betty & Ruth, we used to manage our internal images through a mess of screenshots and desktop folders. We had never thought about using more advanced image search to manage our internal images for design and marketing. Our biggest issue is recalling images ("Where is there a picture of that pink ruffle blouse from last season?").

We started using online storage services, like Box. They have an image-recognition feature, which has made it easier to find things. It's not perfect yet, but it helps a lot.

During our fitting sessions and design processes, we upload process images from our phones. The practice sounds basic, but like many other brands we were using old digital cameras before, and it was difficult to manage.

Neural Networks

Emotions are enmeshed in the neural networks of reason.

—Antonio Damasio, neuroscientist

Neural networks are a type of machine learning model. Chapter 1 provides a brief overview of neural networks, which are modeled after early theories on the way the human brain works. Many methods can be used in a neural network, and nuances in data are being used to train these networks.

Neural networks use example data to infer rules for characterizing new data. The main idea behind neural networks is that you give the system many example answers, and then that collection is analyzed to infer patterns. This is an example of a supervised learning training process. Once the model has been trained, it can analyze new data and label it based on the inferences it made in the training process.

Types of Neural Networks

Neural networks are commonly used, but the architecture of the neural network can have a large impact on have effective they are for a given application. Feed-forward neural networks, recurrent neural networks, and convolutional neural networks each have their strengths. Feed-forward NNs are one of the simplest neural networks. Recurrent NNs are useful when the order of the data is important like in language-based applications. Convolutional NNs were inspired by the human visual system and are often used for imagebased applications.

Feed-Forward Neural Networks

A **feed-forward neural network** is the simplest form of neural network. In feed-forward neural networks, the data being passed through the network travels in only one direction. These algorithms take inputs and then generate outputs. Feed-forward NNs allow signals to travel in one direction, from input to output.

Mentioned in Chapter 1, neural networks are typically organized into three basic layers, though many extend well beyond three layers in practice. These basic layers include an input layer, hidden layer, and output layer, as shown in Figure 4-4.

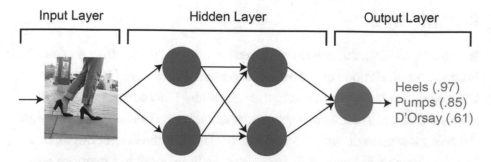

Figure 4-4. *The flow of information, input, and output in a simplified illustration of a neural network. There isn't usually a single input node, but rather a collection of images or other types of data.*

Input Layer

In the first layer, the input layer, no computation is performed. This is the part of the process in which information is passed into the hidden layer. In the example shown in Figure 4-4, the input is an image of a woman from the knees down wearing heels.

Hidden Layer

The hidden layer is where computation is done. Each hidden layer is only a single layer, but there can be more than one hidden layer.

Each input value, in this case an image, is passed through each node in the hidden layer. Each input is given a **weigh,** or **bias**. Weights refer to the strength of the relationship between nodes in the hidden layer.

Neural network training is about figuring out what the weights should be. Each input used in the training process contributes to fine-tuning the weights between nodes. As the network is trained, the weights are adjusted based on the neural network's performance. The performance is evaluated by running the model on labeled data that was not used in training and seeing how well the neural networkpredicts the label.

Output Layer

In the output layer, the **activation function**, also called the *transfer function*, is triggered. Each node in the output layer will return a yes (1) or a no (0) before continuing information to the next node. The activation function does a lot of math to interpret what happened inside the hidden layers and determine what to do with it. In an image classifier, the output might look like the example in Figure 4-4, which shows three possible categories and the probability that those categories are correct, given the image.

Recurrent Neural Networks

Recurrent neural networks (RNNs) are particularly useful when it comes to **sequential data**. Arranging data in order is important for applications like natural language processing and speech recognition. Recurrent networks can have multiple hidden layers. While feed-forward neural networks can also have multiple layers, they allow signal to travel in only one direction, from input to output. Figure 4-5 shows a simple recurrent network cycling through the hidden layer, where the data is processed multiple times using the same function and parameters.

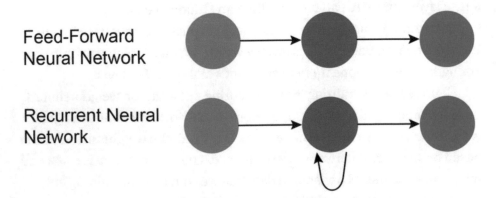

Figure 4-5. *The basic difference between a feed-forward neural network structure and a recurrent neural network*

Convolutional Neural Networks

Convolutional neural networks (CNNs) are better suited for working with images. In CNNs, the neural network will find features in a large dataset and use those to determine what is in the image. The design of CNNs was inspired by the visual cortex system specifically for image-based tasks.

It can be difficult to understand what is going on in the hidden layers of any neural network, but in a CNN, there are two major parts: feature extraction and classification. Features of an image are being detected, narrowing down what is contained in that image. An image with a sweater might be classified by characteristics such as knit structure, the presence of skin, sleeves, and collar. A hypothetical, simplified example of this is shown in Figure 4-6.

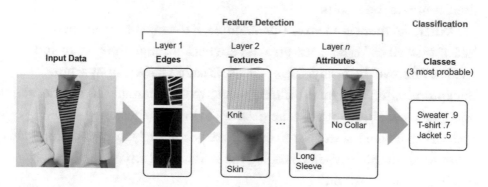

Figure 4-6. *A hypothetical, simplified example of how a CNN extracts features to classify an image of a garment*

A CNN can have tens or hundreds of hidden layers. Each layer can detect different features within an image, increasing complexity with each layer, as in the image shown.

Training Neural Networks

After a neural network is set up, it needs to be trained. **Training** is an important concept in machine learning and is broadly applied to machine learning models, not just neural networks. By sending training data through the model, it "learns" to generate results.

There are two main approaches to training neural networks: supervised and unsupervised learning.

Supervised Learning

In **supervised learning**, the inputs and desired outputs are provided. For example, an image of a dress is used as an input, and the word "dress" is used as an output. (This example would be applicable in classifying large sets of images.) With the already labeled data, the network can process and then compare the results.

Wherever there is an error (for example, if the machine returned "skirt" instead of "dress"), the error is sent back through the system and used to improve the results. This process is called **backpropagation.** Backpropagation compares training results to the manually labeled results and feeds them back through the network to improve accuracy. The system can adjust the weights at each node accordingly to correct the error. Training is complete when changing the weights at each node no longer produces a better result.

This is a common strategy for training neural networks. Supervised training of neural networks relies heavily on the quality of the training data. The networks cannot learn without high-quality, accurately labeled data.

Unsupervised Learning

Another strategy used in training is **unsupervised learning**. In this case, the network is not told the correct or desired output and must decide for itself which features to use to classify data and self-organize. This behavior is commonly called **adaption**. The reason that unsupervised learning is a goal is because there is an ever-increasing amount of easily accessible unlabeled data, whereas creating labeled data can be a time-consuming and costly human task. However, it is a much more challenging approach.

We won't discuss unsupervised learning techniques in great detail in this book, but it's important to know that this is a field of study that has the potential to address a major pain point in machine learning: manually labeling huge datasets.

Training Data

The training process requires large datasets in order for the neural network to find patterns across images. A number of datasets have been created and made publicly available in the research community. While in many industries, information is closed, machine learning has evolved so rapidly and effectively in part because of the sharing of valuable information and resources like these training libraries.

Standardized Datasets

Using standardized datasets helps reduce the variables and isolate problems in designing neural networks and other models. However, standardized datasets introduce other challenges such as perpetuating bias across multiple systems.

One of the most commonly used training datasets is the **Modified National Institute of Standards and Technology** (**MNIST**) database. The MNIST database is a set of 60,000 images of handwritten characters, sampled in Figure 4-7.

Figure 4-7. *Handwritten examples from the MNIST dataset*

Relevantly, a more effective training dataset called **Fashion-MNIST** was introduced by Zalando in 2017. Zalando is a German-based e-commerce retailer that specializes in fashion and beauty products. The Fashion-MNIST dataset contains 60,000 garment images (instead of handwritten characters) as the training data and is said to be more representative of modern computer vision tasks. The dataset offers greater variance and complexity of images compared to the MNIST database. Figure 4-8 shows a sample of images from Fashion-MNIST.

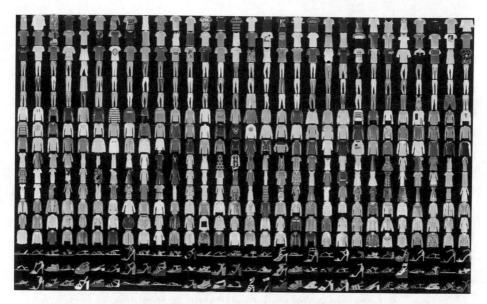

Figure 4-8. A sample from the Fashion-MNIST database

Like the MNIST dataset, Fashion-MNIST contains ten categories of images. In this case, instead of numbers 0–9, they are T-Shirt/Top, Trouser, Pullover, Dress, Coat, Sandals, Shirt, Sneaker, Bag, and Ankle Boots. Rather than the simple features of numbers (lines, curves, and loops), the fashion dataset represents more-complex features (necklines, sleeves, and much more).

In this field, new datasets and tools are made available all the time, improving the work of researchers training new models and applying existing models.

With these improvements, obstacles are also sometimes introduced. Examples of the ways neural networks can be exploited are also sometimes revealed.

Adversarial Examples

In order to be irreplaceable, one must always be different.

—Coco Chanel, fashion designer

While emerging technologies are exciting, they commonly require solving problems that don't exist elsewhere. With any new technology, it's just as important to address weaknesses and threats associated with the new technology. Machine learning does have a long history, but in many cases the research is still nascent, and security exploits are commonly found. While this is arguably true of most of computing, in certain applications, security vulnerabilities could pose a danger to humans.

Adversarial examples are an example of possible security exploits in machine learning. An adversarial example is sort of like an optical illusion for computers that causes the computer to misinterpret something. These examples are intentionally created by an attacker to trick a machine into making a mistake.

Adversarial examples demonstrate that minor changes, which are imperceptible to humans, can dramatically change the results given by a machine. These examples can apply to numerous mediums, particularly images and three-dimensional objects.

Adversarial Image Overlays

There are some well-known adversarial examples. In their paper, "Explaining and Harnessing Adversarial Examples," Ian Goodfellow et al. show an image of a panda that with a few minor changes imperceptible to humans is classified as a gibbon. This example is shown in Figure 4-9. The image in the middle is output by the adversarial example. When overlaid on the image of the panda, the resulting image to the right is incorrectly classified as a gibbon.

"Panda"
57.7% confidence

$+ .007 \times$

$=$

"Gibbon"
99.3% confidence

Figure 4-9. *A small change that is imperceptible to humans can cause a machine learning model to confidently output the wrong label*

Adversarial Additions

Tom Brown et al. in December 2017 published an example of being able to create a unique *adversarial patch* that, when present in an image, prompts the machine to ignore other objects in the image and classify the image as a toaster. This patch is able to trick a machine into believing that the image contains a toaster in a variety of environments. The use of this patch is shown in Figure 4-10.

"Banana"
97% confidence

$+$

"Toaster"
99% confidence

Figure 4-10. *The addition of an adversarial patch to a scene with a banana is able to trick the model into falsely labeling it as a toaster*

Adversarial Objects

As a final example, a 3D object can look like a turtle but be classified as a gun by a machine from every angle. While some of the examples shown using images are not always as robust under transformations like cropping and rotating, 3D adversarial examples are. Anish Athalye et al. demonstrated a reproducible method for creating these 3D examples in their paper, "Synthesizing Robust Adversarial Examples," in October 2017. Their 3D-printed adversarial turtle can be seen in the images in Figure 4-11.

■ classified as turtle   ■ classified as rifle   ■ classified as other

Figure 4-11. *The model classifies images of a turtle as a rifle with high confidence from almost every angle*

Possible Implications

Is there a harmful way in which these exploits can be used in the fashion industry? We might not be able to imagine this possibility yet. We can only guess: what if someone created a neural network for spotting knock-offs? Another person could create images using these examples to reclassify fakes as real designer goods, or vice versa.

A common "what if" scenario that causes alarm in self-driving cars is the threat of stop sign that has been perturbed, or slightly modified to deceive the machine's system. If a computer vision system in a self-driving car does not perceive a stop sign because it has an adversarial sticker placed on it, the car might not stop. These are the kinds of examples that give people pause when thinking about implementing new technologies on a massive scale.

Summary

Images are used frequently in fashion. Historically, discovering these images has been based on text queries. New methods for finding images are emerging because of developments in neural networks. We can now understand much more about the content of images we use every day in fashion.

Neural networks offer a way to automate the process of understanding, "What is this?" in visual data. Convolutional neural networks are especially useful in this scenario. They were designed with the task of processing visual data in mind.

Adversarial examples show vulnerabilities in neural networks and how they can be exploited and manipulated to output the incorrect results. While for some applications this might be less problematic, in high-value or high-risk situations, they can go so far as to pose a threat to human life. It's crucial to consider, what can go wrong if I make an incorrect guess?

Terminology from This Chapter

Activation function—Defines the output of an individual node. In neural networks (NNs), this is also commonly referred to as the *transfer function*.

Adaption—A method in unsupervised learning in which the network decides for itself what features to use to classify data and self-organize.

Adversarial examples—Exploits that are able to confuse a neural network or other machine learning model, thereby getting that model to confidently output the wrong answer.

Alt text—Metadata often used in HTML markup to describe the contents of an image. It is often implemented by the engineer writing the code, and can be important for search engine discovery.

Backpropagation—Uses errors comparing training results to the correct results and feeds them back through the network. The model becomes more accurate by feeding learned information back through the system.

Bias—In neural networks, a bias is a number that represents the strength of the relationship between nodes in the hidden layer. Adjusting the bias, or weight, is a critical part of the training process. See also *weight*.

Convolutional neural networks (CNNs)—A type of neural network designed for interpreting visual data.

Fashion MNIST—The commonly-used successor to the MNIST dataset, Fashion MNIST is composed of 70,000 labeled fashion images.

Feed-forward neural networks—The simplest form of artificial neural networks, in which information flows in one direction only and never goes backward.

HTML code—A standard markup language used for creating web pages and applications to render UI components.

Image search—A more general term referring to the discovery of images based on a search query. Usually, image search refers to a text-based query, and other terms, like reverse image search or visual search, are used to describe image-based methods.

Image tagging— A process of describing what is in an image by using keywords.

Keywords—Also referred to as *index terms*, these capture the essence of an image or document and make it searchable, particularly on the Internet.

Markup—In computer processing, a system for annotating text that modifies how the text is displayed. HTML is a commonly used markup language. A markup language is used to indicate sections of a text document (including headings), images, and styling differences.

Metadata—Data that refers to other pieces of data. Metadata could include information like author, title, description, and location. It may be hidden from the user, but machine readable for discovery.

Modified National Institute of Standards and Technology (MNIST)—The MNIST dataset is commonly used for training and testing image-based machine learning models. It is a databased composed of 70,000 hand-written characters, numbers from 0-9.

Recurrent neural networks (RNNs)—Useful when arranging data in order is important; that is, for natural language processing and speech recognition.

Reverse image search—A search method that uses a user input image to find similar images.

Search engines—Find items like documents or images that are related to a user's input by using keywords. Search engines are used both locally and on the Internet.

Sequential data—Data that requires a special order to make sense. A sentence is an example of information that, taken out of sequence, can lose all meaning.

Supervised learning—A method of learning in which a model is trained on labeled training data.

Training—The process in which a model learns from training data.

Unsupervised learning—Uses unlabeled data to train neural networks and other machine learning models.

Visual search—A search tool for finding objects within an image that are returning results similar to a given object. This process differs from reverse image search in that the results are not related to the image as a whole, but rather related to the objects it contains.

Weight—In neural networks, a weight, or bias, refers to the strength of the relationship between nodes in the hidden layer. See also *bias.*

CHAPTER 5

Virtual Style Assistants

Assistance is the new black.

– Aparna Chennapragada, director of product, Google

The idea of creating an artificially intelligent personal stylist has been frequently revisited in popular entertainment. The first time I can remember being exposed to this idea was watching the film *Clueless* (circa 1995). Cher is choosing her outfit for school by using a computer system that tells her "Mis-Match" for outfits that don't style well together and shows her what the outfit will look like on her.

According to the McKinsey and Business of Fashion report for 2018, "75% of retailers plan to invest in AI over the next two years."

The AI personal stylist concept is a culmination of the technologies discussed in this book, and a look into the future of specialized assistants. The virtual style assistant will be one of the more intimate in its class, knowing more anthropomorphic data about users than other virtual assistants and comparable products. It also provides a personalized future for e-commerce.

The virtual style assistant brings together several areas of artificial intelligence, including natural language processing, natural language understanding, computer vision, neural networks, and other types of machine learning.

© Leanne Luce 2019
L. Luce, *Artificial Intelligence for Fashion*,
https://doi.org/10.1007/978-1-4842-3931-5_5

Virtual Style Assistants

The only real elegance is in the mind; if you've got that, the rest really comes from it.

—Diana Vreeland, former editor-in-chief, *Vogue* magazine

The virtual style assistant is useful when it comes to fashion sales. Bringing personal stylists to the retail and e-commerce environment can improve a brand's ability to match consumers with desired products and guide context-based decision-making. It can also be brought into the home of the consumer, making better use of existing wardrobes. An AI stylist can help consumers discover apparel items that meet a wide variety of expectations: flatter their figure, work well as an outfit, align with current trends and values, and provide a personalized experience.

To understand the impact and context of creating an artificially intelligent style assistant, it helps to be familiar with both the role of a personal stylist and recent developments in technology.

Personal Stylists

Personal stylists help individuals to look their best by curating clothing, outfits, makeup, and other aspects of personal style. This service is often available to only people who can afford it. Hiring a personal stylist would be considered out of reach for average citizens across the United States because of economic or geographic constraints.

Several solutions on the market use technology to address a desire for personal styling. Platforms that act as social networks between stylists and consumers bridge this gap. However, because they continue to rely on services by humans, these services will not scale at the rate that an AI style assistant can. It is because of its scalability and potential to capture value in this field that the AI stylist is a compelling use case.

AI stylists just aren't very good compared to humans right now. Style encompasses many things machines still don't understand. The introduction of an AI stylist doesn't mark the end of the personal stylist as an employment opportunity. Stylists curate aesthetics for their clients, often interpreting between the lines as the clients describe their personal style. They help guide people through an experience that might cause them to feel fear, uncertainty, insecurity, embarrassment, or confusion. Stylists provide a personal experience of shepherding individuals into a journey of self-expression. Humans trust humans more than they trust businesses or software. When it comes to something as deeply personal as our appearances, a real-life human recommendation will always override an AI system recommendation.

Virtual Assistants

Virtual assistants provide the basis for the virtual style assistant and have become increasingly prevalent in consumer electronics. These assistants generically refer to a software agent that provides services to individuals. Today, this is often carried out through voice command prompts. By brand name, virtual assistants refer to Apple's Siri, Google's Google Home and Google Assistant, Amazon's Alexa, and other similar AI-based assistants.

The first step of the process for these systems is interpreting human input. For today's virtual assistants, that usually means converting speech to text. The AI records the human voice and creates text from those recordings in real time. This process is generally referred to as **speech to text**, or **automatic speech recognition** (**ASR**).

Voice Interfaces

The ability to purchase products is also available in many virtual assistants. Voice interfaces, sometimes referred to as **voice user interfaces** (**VUIs**), are changing the way that consumers behave. According to Google/

Peerless reports from August 2017, 58% of people who use a voice-activated speaker are now creating and managing shopping lists at least once per week, and 62% say they are likely to purchase something through their voice-activated device in the next month.

The hardware that is being manufactured today is able to support **far-field voice input processing** (**FFVIP**). This has opened up a wider variety of use cases for voice interface–based devices by making it possible to speak to them from further away. This has been enabled in part by using multiple microphones; the iPhone 5 has three, and the Amazon Echo has seven. The devices use delay between picking up the same sound in different microphones to identify where the sound is coming from and cancel the sound coming into the other speakers.

Features of the Virtual Style Assistant

The virtual style assistant is unique compared to other virtual assistants in that it emphasizes the use of images more than any other use case. Images are critical to giving style advice.

The design of a virtual style assistant requires a few key components: the ability to take a photo of oneself and the ability to store those photos in the application. It also requires underlying technologies such as computer vision capabilities for image recognition and visual search, recommendation engines, analytics, and access to fashion products.

Existing Examples

Images collected by a virtual style assistant can be put to use in several ways. In this and other existing virtual style assistants, images are used to catalog a wardrobe, make product recommendations, and provide insights into the user's style preferences.

Amazon's Echo Look is the most public example of a virtual style assistant. In addition to Amazon, various startups have begun to develop technologies for this purpose. From the behind-the-scenes software created by some of the subscription box companies to unique apps (for example, Lookastic, which makes recommendations based on your response to their "What's in your closet?" interface). Companies such as MemoMi are building smart mirrors that can assist customers in retail locations.

Defining personal style is difficult. In some ways, other platforms have been used to meet the need for a virtual style assistant. Pinterest, for example, recommends images on their web site that are like those you collect on boards. Fashion is one of their most popular types of content. Once images are recommended or discovered by the user, some of them are even shoppable.

Amazon's Echo Look

The Amazon Echo Look was released by Amazon in 2017. The device is intended to operate as a personal stylist, most notably giving AI-generated advice about how two outfits look on you when compared. The Echo Look app provides these software features:

- It allows customers to take better selfies. A one-time setup results in a series of consistent selfies. It provides flash and some image correction to create a focus on the person and the outfit. Better selfies provide more consistent data for Amazon to use when providing analysis and recommendations.

- It allows the customer to create a history of outfits they've worn.

- It recommends garments that you can buy on Amazon based on what you're wearing in a given photograph.

- There are insights about the colors in a customer's wardrobe and their prevalence.

As the history of images of outfits grows, that library is more likely to be useful to the user. The Amazon Echo Look is shown in Figure 5-1.

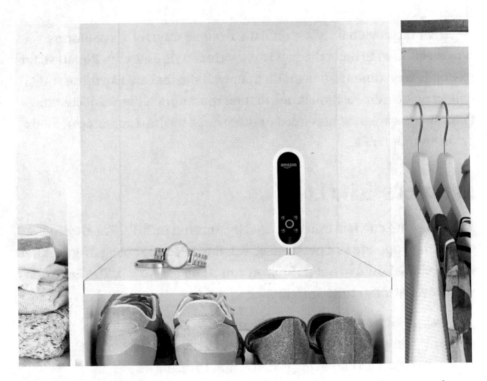

Figure 5-1. *The Amazon Echo Look (image courtesy of Amazon)*

The Hardware

The hardware in this device is packed with features. It is capable of doing far more than the Echo Look application requires of it. The app itself could perform the functions it provides using hardware from your phone. It has a 5-megapixel camera as well as an infrared camera that gives it the ability to sense depth. In theory, it could accurately identify points of measure and measurement data from the bodies it photographs. The device can also store up to 8 GB of data locally. We can only speculate about what Amazon is doing with the data from this extravagant piece of hardware.

Infrared cameras detect heat and produce a thermal image to represent that heat. The Echo Look will blur out the background surrounding the person in the image. It is using infrared to find the human in the frame and blur everything around its figure.

Mobile-device technology is quickly outpacing this approach. Google's introduction of Portrait Mode on its Pixel 2 device released in 2017 features similar technical capabilities, such as putting the subject into focus and blurring the background. In fact, Google's introduction of Google Lens also indirectly competes with Amazon shopping recommendations in the app. Figure 5-2 shows an image of Google Lens finding an exact match for the maroon shirt pictured in the background. In the future, the most likely virtual style assistant will be your phone.

Figure 5-2. *An example search result from Google Lens*

Image-Based Reviews

Image-based reviews provide a huge amount of information to consumers who are making a decision about purchasing a fashion product—including insights to fit, fabrics, and details that might not be obvious in the studio photos found on product pages.

Despite the value of image-based reviews, none of the top ten fashion retailers include them on their e-commerce sites. In fact, of that top ten, half of them don't have customer reviews on their web sites at all. It's rare that people today will go anywhere, do anything, or buy anything without reading the reviews.

In other e-commerce categories, image reviews are commonplace. Of the top three home furnishing retailers—West Elm, Pottery Barn, and Williams-Sonoma—all three feature image reviews prevalently on their site's product pages.

Why in the fashion industry have image-based reviews been so neglected? Today, it's inconvenient to take a picture, find the product page, and upload a review of a garment. The process is messy and time-consuming, too much of an ask for a busy customer.

The Future of Image-Based Reviews

On our smartphones, when we are inside a restaurant, our map app knows we are inside that restaurant. We are more likely to leave reviews as we are prompted to in that moment, or later when we're reminded that we had been there. We might even include an image of the dish we ate.

Over time, it will become easier and more prevalent for our mobile phones to have information about what we're wearing. In the future, they will find the product pages for us and prompt us to leave garment reviews while we still have that garment on. Our mobile accounts will have a history of all the things we've photographed ourselves wearing and may even recommend outfits curated from what's already in our closet, featuring our favorite bloggers, lookbooks, and other users who have shared their outfits.

Artificial General Intelligence

Like, yes—in particular areas, machines have superhuman performance, but in terms of general intelligence, we're not even close to a rat.

—Yann LeCun, head of AI research, Facebook

The virtual style assistant is in some ways aligned with the goal of **artificial general intelligence** (**AGI**) to represent the full spectrum of human intelligence.

AGI is also referred to as **strong AI**, **full AI** or **AI complete**. The specialized intelligence discussed in the rest of this book is often referred to as **weak AI**, **narrow AI**, or **applied AI** because its use is usually limited to a specific application.

Strong AI is rooted in the premise that artificial intelligence should achieve the same levels of intelligence as humans. Currently, this concept is overshadowed by applied AI methods, not because it isn't a desired outcome, but because there is lower-hanging fruit. Strong AI is a really hard problem that hasn't been solved.

Applied AI is focused on specific tasks or areas of problem solving. Even the virtual assistants we are interacting with today embody narrow AI systems. When a user is interacting with a system like Siri, or Google

Assistant asks a question that is out of reach for the system, it returns an Internet search query rather than smartly connecting the user with the app or activity that they are trying to accomplish.

Hybrid Intelligence

To create the illusion of artificial general intelligence, some companies, such as Fin, have created hybrid virtual assistant services. These hybrid systems capitalize on the strengths of current artificial intelligence and use humans to assist, covering the gap between what machines can do and what humans need. These systems are a stepping stone that enables us to learn about the type of assistance humans are looking for and to close the gap between what is possible and what is useful and expected.

Pitfalls of Artificial General Intelligence

Unfortunately, with the rise of more powerful and accessible tools, a lot of misinformation and misunderstanding about AI has emerged in the public media. For some people, oversimplifications have led them to believe that we are further ahead than we actually are in developing AI tools. Some may be disappointed to find that AI falls short of their expectations.

Dangers of AI

Just as we use law to create a system of checks and balances that guide human behavior, we will do the same for AI systems and again for the humans that guide them.

There is lot of media discussing the potential dangers of implementing artificially intelligent systems. For most experts, this caution isn't about an AI machine becoming a sentient being and taking over the world. The warning comes in really practical problems that can arise when relying on AI alone.

AI systems, for example, currently do not understand cultural, social, or ethical norms very well. They generally cannot exercise good judgment in the face of complex scenarios that require a lot of context. In terms of NLP and NLU, we have already seen major disruptions to the US economy and other economies based on real-world reactions of these technologies in financial systems.

In 2008, a news article was accidentally published declaring the bankruptcy of United Airlines. Because high-frequency traders were relying on analysis by NLU, automated trading reacted in a matter of seconds. The market value plummeted, losing $1 billion in a matter of 12 minutes. Examples like these, of which there are several, don't necessarily emphasize the danger of AI. The humans building these systems can't always predict disasters like these but have become more aware of adding safety checks.

Summary

The virtual style assistant is an idea that has provoked our imaginations for decades. Finally, this concept is coming into fruition with the implementation of automatic speech recognition, natural language processing, computer vision, and other technological advances in the last decade.

The virtual style assistant is a specialized concept stemming from the emergence of more-general virtual assistants. This application of AI emphasizes voice interfaces to complete tasks for users at their spoken request. With only a few examples of the virtual style assistant, many of the features for this application remain to be determined, leaving a large opportunity for development in the space.

Terminology from This Chapter

AI complete—The mindset that AI should carry out all human cognitive capabilities. See also *artificial general intelligence (AGI)*, *full AI*, and *strong AI*.

Applied AI—The application of AI to real-world, specific problems, often outperforming humans at these specialized tasks. This is the most prevalent form of AI today.

Artificial general intelligence (AGI)—The goal of creating "thinking machines" that serve as general-purpose systems that are as intelligent as a human. See also *AI complete*.

Automatic speech recognition (ASR)—Turns spoken human language into text in real time. This is a prerequisite task to voice interface systems.

Far-field voice input processing (FFVIP)—The processing of voice commands that are dictated at a distance from the microphone of a smart device.

Full AI—See *AI complete* and *artificial general intelligence (AGI)*.

Narrow AI—See *applied AI*.

Personal stylist—A professional who consults individuals on their personal style, including hair and makeup, fashion and accessories, and other aspects.

Speech to text—Also described as speech recognition, speech-to-text takes spoken language and translates it to text. See also *automatic speech recognition (ASR)*.

Strong AI—See *AI complete* and *artificial general intelligence (AGI)*.

Virtual assistants—Specialized AI-based assistants including Google Assistant, Siri, and Amazon Alexa. See also *AI assistant*.

Voice user interfaces (VUIs)—Allow people to use voice-based interactions to interface with machines rather than using screens.

Weak AI—See *applied AI*.

PART III

Sales

CHAPTER 6

Data Science and Subscription Services

Fashion is part of the daily air and it changes all the time, with all the events. You can even see the approaching of a revolution in clothes. You can see and feel everything in clothes.

—Diana Vreeland, former editor, *Vogue* magazine

Data is shaping the way we experience retail by enabling customized experiences. In the fashion industry, the technology being built around subscription services provides an example of how these custom experiences can be applied to e-commerce.

While for most fashion brands data is important for driving sales and producing styles that customers want, subscription services often rely on data-driven custom curation as the only method of product discovery. There may be no online catalog or search bar to make purchases from these businesses. Because of this, subscription services have a higher cost of failure and are deeply motivated to develop methods to deliver exactly what a customer wants on an individual basis. Using data to keep customers captivated has become part of the DNA for companies like Stitch Fix, Rocksbox, and Le Tote, just to name a few.

© Leanne Luce 2019
L. Luce, *Artificial Intelligence for Fashion*,
https://doi.org/10.1007/978-1-4842-3931-5_6

Subscription Models

Women could have a closet on rotation, have unlimited possibilities of what to wear...

—Jennifer Hyman, CEO, Rent the Runway

Numerous mechanisms are being used to build subscription-based business models. Even before diving into the details of how customization and data are being applied, this type of business is new to many industries and requires a bit of a backstory.

Subscription business models can take many forms. Some might be pay-as-you-go, while others might have an annual fee that's paid in one lump sum. Services offer customers varying degrees of choice and surprise in what they receive. Customers might be totally surprised every time, or they might select some of the products they want in advance.

In many subscription services, information is collected about which items are accepted or rejected in a given shipment. The user is usually prompted to answer a brief survey about why they rejected a given garment. They could answer quantitative questions such as how they would rate their satisfaction on a scale of 1–10, or they could answer questions with text that give them an opportunity to express themselves. Specialized techniques in natural language processing and machine learning can help analyze and quantify free-text responses to learn from them at scale. With insights that address specific areas of satisfaction or dissatisfaction, these companies grow an understanding of which garments and which customers are well matched and various other information about both the product and the customer.

In these businesses, brands can learn about what customers think of a given product. They might collect responses from thousands that uncover problems with the garment, shown in Figure 6-1.

Garment   **Survey**   **Users**

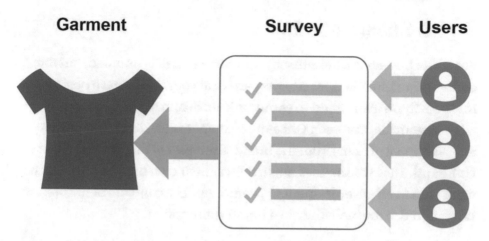

Figure 6-1. *The network of users giving feedback about a garment*

They also collect data about the user's preferences based on the garments they select, shown in Figure 6-2.

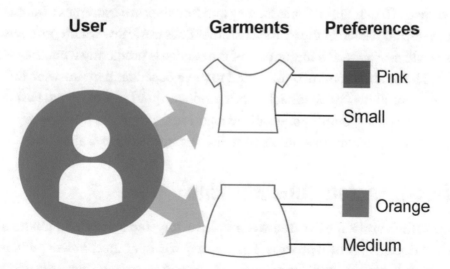

Figure 6-2. *The user's preferences can be "learned" through the choices the user makes about which items to keep and what they say about those items in a survey*

Brand Subscriptions

For some businesses, the subscription service itself is branded, but the contents are kept a secret until they arrive at your door. These businesses rely heavily on their brand to communicate that their products are aligned with the customer. Once the customer subscribes, they receive an assortment of items from the brand. Examples of this are Causebox or Unbound. These businesses might deliver their own brand or brands they work with as an advertisement or promotion. The contents of the box are not tailored to the individual in a brand subscription.

Targeted Subscriptions

Other subscriptions ask customers a series of questions that help them pinpoint exactly what kinds of products they're looking for. This targeted subscription sometimes takes the form of a personal stylist service, as in the case of Stitch Fix or Dia&Co. Ipsy and Birchbox are examples of similar businesses. In each of these, the customer does not know which products they will receive, and a major part of the service is predicting what they will like. The products are recommended to them based on their answers to survey questions and feedback about products they have received. Fashion services typically employ specialized algorithms alongside humans to bring the experience of a personal stylist to many users at scale.

User-Selected Subscriptions

Trunk Club and ShoeDazzle are examples of targeted subscriptions with a user-selected approval process. The stylist or a similar mechanism provides options to the user, and if the user confirms that they want them, the items are delivered. At that point, the user still has a choice to either purchase or return the items. This often comes with a styling fee, which likely is used to cover two-way shipping costs on orders when no purchase was made.

Consumables Subscriptions

Dollar Shave Club and Billie, a women's version of a similar service, both fall into a consumables category of subscription services. The user receives the same product at a set frequency—once per month, for example. These subscriptions allow people to automate the procurement of items they use every day.

What we culturally consider "consumables" has evolved over time. Items like socks and underwear, which are replaced more frequently than other apparel items, might fall into this category. Consumer habits around fast fashion have turned fashion items into consumables as well, creating new opportunities for subscription services like Stitch Fix.

Rental Subscriptions

The concept of renting items in the fashion space is not new. Suits and tuxedos, historically expensive purchases for men, have been widely available for rent since the 1970s. Rent the Runway got started by bringing this concept to women's fashion. They have since introduced subscription services around their rental garments, making it possible to rotate items in and out of your wardrobe every month.

What's different about a rental subscription is that it is recurring. These services allow the user to choose what they'd like to rent and ship it to them for a monthly fee. There are usually some rules around the duration of the rental and other constraints to the service. Unlike many other subscriptions, this model also allows the user to choose what they want.

Rent the Runway has created a unique community of women sharing a wardrobe with each other. They are incentivized to also behave in ways that are dissimilar to other e-commerce platforms. Rent the Runways customers are willing and excited to share images of themselves in the garments even when they look bad, provide data about their body types and the fit of the garments, and explain the wearing occasion.

The engagement of these customers allows Rent the Runway to constantly learn about items that are popular and the many properties that make those garments desirable. The information allows other customers to make more informed opinions about their rentals. Even negative experiences end up being valuable to everyone in the process.

Digital Personalization

Subscription services, especially targeted subscriptions, are important because they are able to embrace the benefits of digital commerce through personalization.

The emergence of digital products has made personalization an expected experience for consumers. They interact with digital products that remember what they watched and liked and that infer preferences they didn't even know they had. As the amount and accessibility of information and content has grown quickly over the past several decades, this type of personalization has gone from being a convenience to a necessity.

Hyperpersonalized recommendations for music, movies, digital products, and other digital services have become expected. Netflix, for example, doesn't show you every movie in existence when you log in. Instead, Netflix makes personalized recommendations based on other movies you've watched and specific interests relevant to your profile. As this has become the new normal, consumers are demanding it in every aspect of their lives. It gives them a sense of control and reduces the paralysis of choice. All of these experiences are driven by collecting data about what that user is doing on the site and then turning it into actionable customization.

For a consumer, going back into the fashion retail environment after being exposed to this level of personalization is like entering a world of complete and utter chaos. Suddenly, someone who is used to doing very little to get to a product they want has to sift through hundreds or thousands of unwanted items.

Discovery and curation is changing in a number of ways. It's shifting away from the fashion brands and retailers to new categories of sales channels. Online marketplaces like Amazon and social media marketing platforms like Instagram are improving their ability to recommend products to users. The methods they use to generate recommendations are based on some foundational concepts: recommendation engines, databases, and statistics.

Recommendation Engines

On a high level, services like Stitch Fix built their entire service around mastery of **recommendation engines**, also known as **recommender systems**. Recommendation engines help users filter out tons of items that users don't want. They're used in virtually every popular service to reduce the burden of choice, from Instagram ads to Amazon product suggestions. Recommendation engines increase the chance of a conversion by returning the right results to a user.

Recommendation engines aren't new at all, and in fact they have their roots in e-commerce. The two most commonly used types of recommendation engines are collaborative filtering and content filtering. A third, hybrid filtering, is a combination of both methods.

With the amount of data consumers are exposed to online, these engines need to be able to learn from and adapt to taking in new information from the user in real time.

Collaborative Filtering

Collaborative filtering uses information from a large dataset of user purchases and other behavior to predict what another customer is looking for. There are two basic approaches to collaborative filtering: a user-based approach and an item-based approach.

In the user-based approach, collaborative filtering returns recommendations based on a user's similarity to other users, as shown in Figure 6-3. Examples of this include product recommendations based on what other users have purchased, which you can find on e-commerce sites like Amazon.

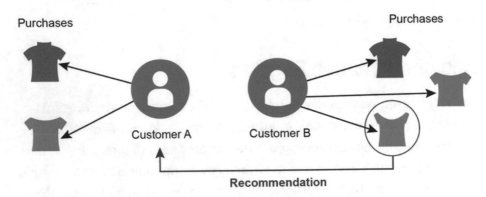

Figure 6-3. *Information about similar customers can be used to recommend products*

An item-based approach recommends products based on how users rated similar items. Item-based collaborative filtering is useful because it can give relevant recommendations even if it doesn't know anything about a given user.

Content-Based Filtering

Content-based filtering methods are based on user actions and preferences. If a user is exploring a site and likes and buys only red dresses, then more red dresses will be revealed to them during their search. However, one problem with this method is that it may continue to recommend products in the same category, potentially even after the user has lost interest in searching for items in that category. An example of content filtering is shown in Figure 6-4.

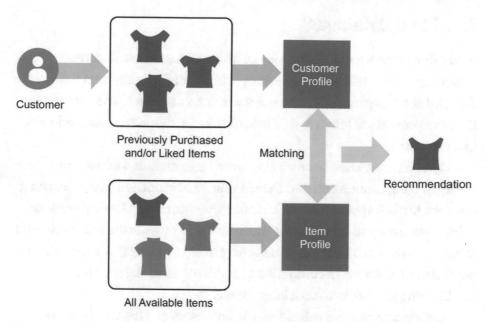

Figure 6-4. *A customer who made a purchase in the past being recommended similar items*

Data Science

Data is a precious thing and will last longer than the systems themselves.

—Tim Berners-Lee, inventor of the World Wide Web

Data science is a multidisciplinary field that uses statistics and software engineering to extract business and product-relevant insights from data. Data science does not necessarily rely on machine learning, but it's becoming more and more relevant as a tool to people working in this field.

Data is often a raw set of information that requires some processing in order to have meaning. Digital data usually exists in databases, which can be structured or unstructured in nature.

What Is a Database?

A **database** gives a machine a way to know where to recall certain information. For instance, a product catalog might contain an item like a skirt, which has properties of color variations and sizes. To know where the size property is located, that information is made machine-readable in a database.

Databases are used to efficiently store large amounts of data and perform computations on that data. A database is not a machine learning concept but is important for all kinds of programming. Databases make information easy to access using machines and provide the foundation for structured data, which is data that is organized in a way that a machine can understand it. This is in comparison to unstructured data, which is more challenging to work with on a larger scale.

A basic database is a called **flat-file database**. In a flat-file database, the data exists in a single set of related data records called a **table**. A flat-file database is a lot like a spreadsheet. In a spreadsheet, you can do things like reorganize rows based on various sorting mechanisms. You can also set columns to contain specific kinds of data or apply mathematical equations to data in that column. Like spreadsheets, database tables also contain rows and columns, but there are a few important differences. In a table, columns have a set **data type** that describes what kind of data the column contains, such as a number or string of characters. The **schema** is the set of rules that defines the table. Though there may be more rules ascribed to a database table, the basic mechanisms are very similar to a spreadsheet.

In a flat-file database, each **record** in the database has its own row in the table. Each column stores some piece of data about the record. A flat-file database could be implemented as a text file that contains one row per line and has data from each column separated by commas. Figure 6-5 shows an example in which each SKU has multiple attributes associated with it. For example, "boyfriend jeans" has a category of "pants" and season "fall winter 2018."

SKUs ◄——— Name

id	style_description	category	season
1	silk bomber jacket	outerwear	ss_18
2	boyfriend jeans	pants	fw_18
3	slouchy tee	knit tops	ss_18
4	l/s button down	woven tops	fw_18
5	denim jacket	outerwear	ss_18
6	cardigan	sweaters	fw_18
7	fit and flare dress	dresses	ss_18
8	tank top	knit tops	ss_18

Attributes

Figure 6-5. *A visual representation of a flat-file database*

If you were to list a brand's entire repertoire this way, your database would quickly grow large and difficult to manage. **Relational databases** are useful for managing clean data by linking related tables together. Rather than having redundancies in the table entries, there can be a separate linked table with more detailed information about that attribute.

Figure 6-6 shows how you might use a relational database to define seasons. In its own table named "Season," each season can be assigned to other attributes such as "delivery dates."

SKUs

id	style_description	category	season
1	silk bomber jacket	outerwear	ss_18
2	boyfriend jeans	pants	fw_18
3	slouchy tee	knit tops	ss_18
4	l/s button down	woven tops	fw_18
5	denim jacket	outerwear	ss_18
6	cardigan	sweaters	fw_18
7	fit and flare dress	dresses	ss_18
8	tank top	knit tops	ss_18

Season

id	season	delivery_date
1	ss_18	03-15-2018
2	fw_18	10-19-2018
3	ss_19	03-17-2019
4	fw_19	10-14-2019

Figure 6-6. *A visual representation of a relational database*

Structured and Unstructured Data

Structured data refers to data that is organized. When data is organized, machines are more easily able to parse that data and make sense of it.

On the other hand, **unstructured data** is unorganized. Generally, this is more difficult for a machine to understand. This is, as mentioned, things like blog posts, e-mails, text messages, and other types of **free-form data**. As we make advances in natural language processing, there is more opportunity for machines to understand human languages. From this, we're able to posit both the content of messages written and the context that makes it relevant.

A good example of NLP in action is using Gmail. Google uses NLP to parse your inbox for information about flights and sends you reminders that you have an upcoming flight.

Data from Words

In Chapter 2, several natural language processing concepts were introduced, including sentiment analysis and word vectors. These techniques can be used to analyze natural language–based reviews, feedback, or comments left by customers. This kind of data is unstructured data; a machine cannot immediately understand what it means.

To understand natural language using NLP, some engineers use word vectors. Investing in techniques to do this analysis means learning and processing more information about your customer than you would have been able to otherwise.

It requires more than just word vectors to make inferences about what customers write, but it is possible to interpret complex phrases that are relevant to fashion businesses. For example, if a customer writes that they are in their "first trimester," you can infer that they're pregnant. If they write that they're "going on their honeymoon," you might infer that they're about to take a trip. It might be obvious to a human, but to a machine, it takes training. This training can be used to help create tailored recommendations for the end customer.

Even if your business is not a subscription service, you can use this information to segment your customer base. With specific details about a customer's needs, left from their comments and feedback about products in the subscription model, you can build more effective ad targeting, e-mail campaigns, and other marketing materials. You can also adopt these insights to shape your approach to product development.

Applications

To take a step back, what would the motive be to understanding the context of words around a product? Let's look at a hypothetical product entry from a fashion brand web site. In this case, it's this image of a tan dress in Figure 6-7. The product in this image is broken into six parts: image, title, description, price, variants, and reviews. This information is limited, because without using techniques to understand the content of the six parts outlined, we are limited to tags, categories, and specific variants listed to narrow our search.

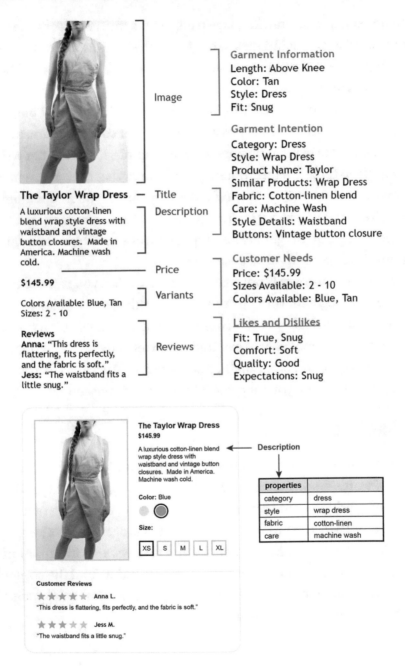

Figure 6-7. Information can be extracted from an example product entry

Each section gives different information about the product that can be used both for search and for product recommendations. Using computer vision, the image can be analyzed to provide more information about the garment and its fit. The product title and description express garment intention—or, more accurately, what the brand intended the garment to be like. The price and variants of the garment give more information about whether that garment suits the customer's needs. Is it available in their size? Are they looking for that color? Is it a price they're willing to spend? Product reviews provide information about how that garment is perceived in the real world. Does it fit the customers who bought it? What does it feel like? Is the quality what they expected?

This maps to a large matrix of characteristics that could be pulled out in a structured way to help pair a person with a product that suits them, as pictured in Figure 6-8.

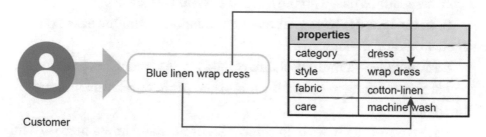

Figure 6-8. *Product characteristics as correlated with an example customer's needs*

Think of search and recommendations as a matching game. How can we learn just enough information about what someone is looking for to deliver the right product to the right person at the time they're looking for it?

Summary

It is not fashion brands, but technology companies in this arena that are taking advantage of the insights data can provide about customers and turning them into personalized recommendations in an otherwise oversaturated market. Ironically, by building businesses this way, these services have often also found that it's more profitable and practical to create their own apparel brands. As they enter into the manufacturing realm, they're taking over market share in the fashion industry. In 2017 alone, Stitch Fix hovered somewhere around 1% market share.

Terminology from This Chapter

Collaborative filtering—A recommendation technique that makes recommendations based on the behavior of similar users.

Content-based filtering—A recommendation technique based on user actions and preferences.

Data science—A multidisciplinary field that uses statistics and software engineering to extract business- and product-relevant insights from data.

Data type—Data types help computers understand how a piece of data will be used by in a program. For example, strings contain words, text, and numbers as human language, whereas integers contain whole numbers that could be used for calculations.

Databases—A structured format which gives a machine a way to know where information is stored making it easier to recall as needed.

Flat-file databases—Store data in a plain-text file.

Free-form data—Unstructured text that is without a format. See also *unstructured data.*

Recommendation engines—A tool used to predict items that a customer or user might like.

Recommender systems—See *recommendation engines.*

Record—The basic unit of entry in a database. For example, your name, social security number, and birthday could be a record in a US Citizens database.

Relational databases—A collection of data broken into well-described tables. These can be challenging to set up but provide well-structured data.

Schema—Describes the data contained in a table of a database. Database schemas can be thought of as the blueprint for setting up the database.

Structured data—Data that is organized in a structured format, like a database.

Table—A structure in a database with columns and rows that holds records.

Unstructured data—Data that doesn't have a predefined data model or organizational structure. It is usually text heavy.

CHAPTER 7

Predictive Analytics and Size Recommendations

I am more than my measurements.

—Ashley Graham, model

Fit is a vaguely defined, complexly intertwined technical and emotional topic. Each individual has a different definition of how they want their clothing to fit. The way that we use words to describe fit even varies from one person to another. "Baggy" to one person may look like something different to someone else.

The Fit Problem

Sizing systems are an alphanumeric organization of garment dimensions created to help individuals find garments that fit. Unfortunately, not all brands assign the same measurements to the same sizes. This means for customers, it can be difficult to figure out which size to buy.

© Leanne Luce 2019
L. Luce, *Artificial Intelligence for Fashion*,
https://doi.org/10.1007/978-1-4842-3931-5_7

According to a Body Labs 2016 retail survey, $62.4 billion worth of global apparel and footwear are returned annually because of incorrect sizing or fit. In these cases, the problems the customer experienced with the fit of the product meant that they were unsatisfied with the purchase and returned it. This was especially prevalent in e-commerce transactions.

All bodies are different. For brands, it is well-known that trying to create products that will fit anyone is fairly unrealistic, maybe impossible. While getting the perfect fit would require tailoring and hands-on personal attention, matching a customer with the best possible garment using historical data can still help reach higher levels of customer satisfaction and decrease returns. This process of fit matching using predictive analytics has shown promise in returning higher levels of customer satisfaction.

What Are Predictive Analytics?

The term **predictive analytics** encompasses a pool of techniques, from statistics to machine learning. The key characteristics are the use of historical data to predict future events.

Humans are really good at recognizing patterns. Having a "hunch" or "intuition" for something is generally more than just a dark art. Predictive analytics are modeled after the idea that by recognizing patterns in events that happened in the past, we can develop a framework, or model, by which we can predict events happening in the future.

A **model** in predictive analytics, as in other machine learning techniques, is a predictive algorithm used to simulate real-world activity. A predictive algorithm is a statistical method that uses historical data to infer what the data might be in the future. The use of a predictive model has two distinct stages: training and predicting, as shown in Figure 7-1.

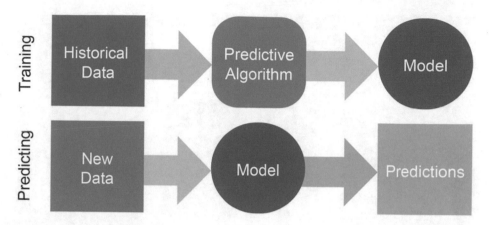

Figure 7-1. *The process of training a model and then using it to make predictions about future events*

Learning Fit

Traditionally, size recommendations have been made on e-commerce platforms through a static sizing chart with dimensions such as waist, hip, and bust measurements. These charts, especially on retail platforms that sell multiple brands using different measurements, have lacked the accuracy that customers need to purchase garments confidently.

Size recommendations apply predictive analytics to match a customer with the size that will fit them for a specific garment.

The way predictive analytics can be applied is by creating a recommendation engine, like those explained in Chapter 6, that provides recommendations based on fit. Recommendation engines are an application of predictive analytics.

A customer might experience size recommendations in an online store by going through these steps: they open an e-commerce site and choose a garment to purchase. If the customer doesn't know their size, they can use the interface to gain insights about what other customers with similar body sizes purchased, as shown the screenshot in Figure 7-2.

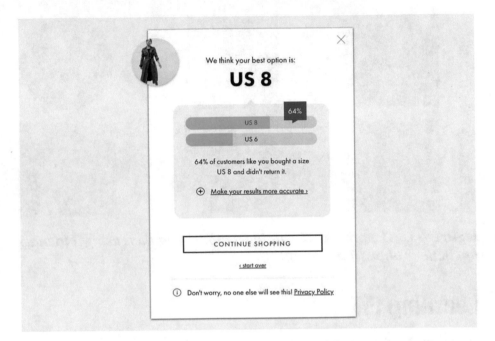

Figure 7-2. *The Fit Analytics Fit Finder on the ASOS web site*

Fit Analytics is one company that provides fit matching and other insights as a service. It powers over 250 million size recommendations every month. Its Fit Finder interface can be found on the e-commerce sites of brands like The North Face, ASOS, and Tommy Hilfiger. After the customer inputs information about themselves (height, weight, age, and fit preference), Fit Finder will return a best-fit recommendation like the one in Figure 7-2. While this concept was new just a few years ago, it has quickly become the new normal on fashion e-commerce sites.

Other Applications for Predictive Analytics

For many companies, using predictive analytics is not a new idea. It can be used across many departments within a business, from sales to marketing to design. Here are a few other scenarios:

- Predicting which customers will continue making purchases from your business and which are likely to leave the platform

- Targeting marketing campaigns to those who are most likely to make purchases

- Identifying suspicious transactions and detecting fraud

Implementing Predictive Analytics Systems

Depending on the business and strategy, a fashion brand can implement predictive analytics by outsourcing and using third-party services like Fit Analytics, described previously. They can also start their own predictive analytics initiatives internally.

For brands that do want to implement their own predictive analytics practices, there are basic steps to take to start a project from scratch. Implementing predictive analytics is done in a series of well-defined stages: define the problem, collect data, create a predictive model, train the model, and then use the model to create predictions. We'll explore each of these steps in the next section.

The most important thing to do is to identify the business value in investing internal time and resources in a predictive analytics project. Without a clear goal, there are too many directions to go and not enough information to learn. To turn to a team and say, "Add predictive analytics to our web site" would be like saying. "Add thread to our clothes." It's too vague and poorly defined to be actionable. Immediately, they will turn to you and ask a bunch of questions about the thread, the business value, and the application: What kind of thread? Who's the vendor? Is it cheaper? Is it stronger? Can the factory use that thread? Which product line? Which category? Which color? Do you want to add that thread to all of our products?

If your role is management at one of these brands and you're looking to start a predictive analytics project, take a step back and evaluate your biggest pain points before diving in.

BETTY & RUTH WORKING WITH E-COMMERCE CLOTHING REVIEWS

At Betty & Ruth, we want to know whether garment size is correlated with customer satisfaction for that garment. In this problem, the input variable is the person's size, and the output is a rating of how likely that person is to be satisfied with the garment.

We created a hypothesis before beginning. We notice a pattern that correlated certain sizes and styles. We believe that we can use customer review data to determine customer satisfaction for a given size per style and make recommendations accordingly. For example, based on reviews for size Medium that say that the style fits small, we can make predictions about how other size Medium customers will respond and recommend sizing up through our web site.

Another benefit to these findings could be that the technical design team learns more about the customers they're developing product for. The information from this experiment might influence their grade rules and technical specs during the product development process.

Data Collection

Once there is a clear goal to the project, collecting data is the next step. No matter the company, you're probably collecting data about your customers and products all the time. **Data collection** is simply about getting the information you need to apply a predictive model.

How to Get Data

For experimentation, Betty & Ruth uses **Kaggle**, which can be a valuable resource for finding datasets. In many cases, these datasets are already prepared for use. Kaggle is a web site where people share and compete on data science challenges. There's a specific focus on predictive modeling and analytics. On this platform, companies and users will often upload clean datasets and challenges. This allows the data scientists on the platform to skip the time-consuming part of collecting and cleaning data and jump straight to writing code to analyze that data.

When we're working with our own data, there are often ways to export information such as customer reviews as a CSV. CSV stands for comma-separated values, and this type of document stores information in a table format by giving each record its own line. If that record contains more than one field, it is separated by a comma. CSVs are a very common file format for importing data.

Some e-commerce platforms, like Shopify and Magento, have a feature built into their Product Review apps for exporting reviews as a CSV file. Even for platforms that don't offer an easy way to do this, there are other resources you can leverage to get it. For example, several web sites and browser extensions will allow you to save reviews from certain e-commerce sites as a CSV.

Cleaning Data

For datasets that is aren't already cleaned and prepared, cleaning data is a vital step to getting quality results. **Cleaning data** refers to the process of removing outliers, spikes, missing data, and anomalies. Data with a lot of outliers is often described as **noisy** because the abundance of irrelevant data makes it harder to interpret the core information. The noise can significantly and unnecessarily change the output of your predictive model.

It might sound inconceivable that data would need to be "cleaned" to process, but it's rare that you would have access to a perfect dataset related to your product and customers. Most experts in this space will tell you that it's better to have clean data than a lot of data. This is true for all applications of machine learning. By some reports, data scientists can expect to spend up to 80% of their time cleaning and preparing data, depending on the project and its data sources.

When data is brought together from multiple platforms (such as Amazon and Shopify, for example), the structures might be different and the fields may not match up perfectly. This is another common reason for needing to clean data. Multiple sources can lead to discrepancies.

Missing data can also be the result of changes over time. On the Betty & Ruth e-commerce web site, we recently started listing fiber contents on every new product. All of the products before that change did not contain the fiber content data. Extracting statistics like "25% of our garments are made out of cotton" without first removing the garments that did not list fiber content, would mess up our statistics. That statement would not be an accurate representation of our product offering. If missing data is ignored or overlooked, which is easy to do (especially in large datasets), the output predictions will be wrong.

Another reason that a dataset needs to be cleaned before being used is because of the presence of **outliers**. Maybe a person who is otherwise a size 0 orders a sweatshirt that is a size XL and thinks that the fit is very snug. This is the kind of outlier that should be removed from your data because it will probably not be predictive of the behavior of the majority of customers. Outliers can be created through human error or machine error and sometimes won't even make sense, given the variables.

Without a high-quality dataset, it is not possible to get high-quality results.

Data Visualization

There are a lot of existing ways for developers and data scientists to visualize data even before jumping in with machine learning and other predictive analytics. By using visualization tools, they can uncover trends in the data early on. Some example programs for visualizing data are Tableau, Chartio, Plotly, Infogram, and Google Charts. Creating these **data visualizations** is helpful for understanding it by giving a visual index of the data.

In a screenshot from Kaggle, you can see how representing a variable in a simple bar graph gives more insight about the dataset than looking at a list view of the same information. Figure 7-3 shows the age distribution of this particular example.

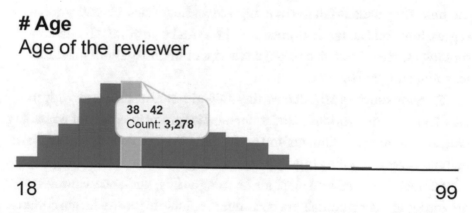

Figure 7-3. *A histogram based on the age variable of an example dataset*

It isn't always obvious at first what type of model you'll need to build. Some exploration and experimentation will have to be done to find useful insights about the dataset and establish more informed hypotheses before getting started on a machine learning project.

Models

In a few paragraphs, it is impossible to explain the challenges that could arise in the actual building and implementation of a predictive model. Some use cases will have easier outcomes to predict than others. Not all predictive analytics models require machine learning, but it is a useful tool, especially in dealing with large amounts of data.

Fortunately, we probably don't need to build a predictive algorithm ourselves to get started using predictive analytics. There are tools available that can be used without even needing to code. They give users access to out-of-the-box algorithms that work for a wide array of problems. Out-of-the-box techniques are useful for getting started, but in practice may not give you the best result.

There are some common algorithms that can be implemented out of the box. They usually fall into two high-level categories: classification and regression. You can learn more about the actual algorithms themselves on blogs such as `towardsdatascience.com` or in books about machine learning and data science.

Platforms such as MLJAR have these algorithms available and ready to use. The platform provides a straightforward graphical interface for uploading data and running experiments and predictions. While it does automate many of the processes required to get up and running, the tools are built for people with domain knowledge in machine learning as their audience. However, a growing number of tutorials are becoming available to those who are curious and want to learn more and experiment without learning to code.

Enterprise Tools

Companies have built graphical user interfaces to make experimentation with data science easier than ever to implement. These interfaces by no means take the work out of implementation, but they make it easier to get started quickly without needing to use **command-line interfaces (CLIs)**, which can be daunting for nonprogrammers. Companies like MLJAR are making interfaces so that as long as you understand core concepts, you can get started. Your data science team might use these tools to streamline their process.

There are also companies, such as DataRobot and RapidMiner, that are offering artificial intelligence for enterprise. These tools usually have an even further developed GUI, making it simpler to use. While they might have less transparency into what's going on behind the scenes, they can help solve large and common problems that might be expensive and specialized to tackle independently—like fraud detection, for example. For more information about the growing number of AI enterprise tools, refer to Chapter 12.

Technology Blogs at Fashion Companies

Several methodologies can be used to match the measurements of a person and a garment. While this chapter does focus on predictive analytics as a methodology particularly for fit evaluation, it's important to mention that variations on this method have also been explored within the fashion industry.

Stitch Fix uses an interesting approach to solve for fit, which is described on its engineering blog. In one example, for each garment, there are three "answers": it's too small, it's too large, and it fits just right. If a customer describes themselves as a size 4 but say a size 4 garment is too large, that garment may be ranked as an in-between size, such as 2-4, and vice versa. Both the user and the garments are being ranked on a measurable scale. Both variables are moving targets, and the system is continually adapted based on information gathered about the user.

Software companies in the fashion space often share their techniques, approaches, and experiments publicly. You can read more about Stitch Fix's approach on its technology blog, `multithreaded.stitchfix.com`.

Stitch Fix isn't the only company writing online about how it is solving problems such as fit in fashion. With a small amount of effort, you can find information about the inner technology at a lot of large software-based fashion companies. A few more examples are Lyst (you can read about its tech at `making.lyst.com`) and Rent the Runway (whose engineering blog can be read at `dresscode.renttherunway.com`). These blogs can provide a source of inspiration and information for other companies looking to update their practices or simply gain insight to some of the mechanics behind these companies.

Data Responsibility

This section might seem a little out of place for this chapter, but when dealing with personal information, companies have a responsibility to protect their users. Especially in recent times, awareness over privacy and security are becoming more prevalent among consumers and corporations alike.

Recent events in this area are driving companies to take security and privacy more seriously. Several major Internet security breaches have left millions of people exposed; for example, the Equifax hack in 2017 exposed the social security numbers of hundreds of millions of Americans. These situations pose a crisis for both financial theft and identity theft. There have also been efforts toward improving privacy and security for Internet users to combat them.

General Data Protection Regulation

In 2018, the European Union (EU) introduced a new policy called the **General Data Protection Regulation (GDPR)**. You probably heard about it, because you would have received an e-mail from almost every account you had, letting you know about privacy policy updates. While it applies only to businesses that have European customers, it has forced a lot of large technology companies to make changes across their business. It covers a wide range of measures to protect user data by keeping it encrypted and private and to give users the ability to opt out of having an account and have their data permanently deleted. It is most strict about regulation around personally identifiable information (PII), which could put the physical safety and privacy of individuals at risk.

Data and Third-Party Vendors

"No one's thinking of us; we don't matter" is no longer a valid way of thinking about security and privacy for businesses on the Internet. That approach leaves businesses more vulnerable to attack because they are low-hanging fruit. It also increases the likelihood of vulnerabilities down the line, when the business does gain traction or publicity.

Using third-party services is common practice, especially for fashion brands whose core competency isn't technology. What those vendors do with customer data is still the brands' business. Consumers are becoming more vigilant and expect at least basic levels of privacy and security. The downside to ignoring the security practices of third-party vendors is that if they get hacked, it could reveal sensitive customer data and destroy customer trust in the brands partnering with them.

Legal

For businesses using recommendation systems, ad retargeting, and other marketing tracking strategies that use user data, their Terms of Service and Privacy Policy should reflect how the data will be used.

Summary

Predictive analytics can provide powerful tools for fashion businesses, and the applications are endless. The most important aspects of implementing an effective and accurate model are having high-quality data and a clear business goal.

With the introduction of new graphical user interfaces for data science and predictive analytics tools, it's becoming easier for businesses to implement powerful machine learning methods.

While implementing predictive analytics and collecting and analyzing data about users, it's a corporation's responsibility to be mindful of data privacy and security. It will prevent financial and personal harm for both the corporations and their customers.

Terminology in This Chapter

Cleaning data—The process of removing outliers, spikes, missing data, and anomalies. Without clean data, predictive analytics models run the risk of inaccuracy.

Command-line interfaces (CLIs)—On every computer, there's an interface, also called a *shell*. The CLI is not commonly used by a general computer user, because most users prefer a graphical user interface. However, for users who do a lot of programming it can provide a powerful interface for controlling programs and the operating system of the machine.

Data collection—The process of gathering information for a specific use. Often in the post-collection process, that data might be mined and cleaned. (See Chapter 9 for more information about data mining.)

Data visualizations—Charts and other graphics that help humans understand the information embedded within a set of data.

General Data Protection Regulation (GDPR)—A data privacy regulation passed in 2018 by the European Union to protect the rights and data of Internet users.

Kaggle—A platform for predictive modeling and analytics challenges that was acquired by Google in 2017. Corporations and individuals have hosted challenges on the platform in which they prepare the data and describe a problem; then data scientists can compete to prepare the best model.

Missing data—Information gaps in a dataset. Missing data is important to be aware of because it can impact the quality of the output in models applied to the dataset.

Model—A mathematical representation used to simulate real-world activity. A machine learning model refers to the resulting artifact of a trained machine learning algorithm.

Noisy data—Data with spikes, outliers, anomalies, and missing data. This is often the state of raw data before it has been cleaned.

Outlier—Exceptions in data. When something is far off from what is normal, it could be because of machine error, human error, or incorrect interpretation. If a dataset reports a human user to be 300 years old, but everyone else in the dataset is within a more feasible age of 25–55, we might assume that something went wrong with that data. It is an anomaly, and we can remove it from our dataset in order to prevent inaccuracies.

Predictive analytics—A wide range of topics and methods, from statistics to machine learning, that use historical data to predict an event in the future.

PART IV

Designing

CHAPTER 8

Generative Models as Fashion Designers

The great thing about fashion is that it always looks forward.

—Oscar de la Renta

Amazon made headlines in 2017 for a controversial idea that it introduced to the public consciousness. Amazon claimed the ability to train a **generative adversarial network (GAN)**, a type of **generative model**, to design garments. For many fashion industry professionals, this announcement set off alarms. The threat that the role of the fashion designer would soon become obsolete hit close to home for everyone.

AI Fashion Designer

Fashion has always been a repetition of ideas, but what makes it new is the way you put it together.

—Carolina Herrera, fashion designer

The use of a generative model as a fashion designer refers to a process of taking in a dataset of images, and outputting images that are visually similar but generated by the model. The images that make up the input

L. Luce, *Artificial Intelligence for Fashion*,
https://doi.org/10.1007/978-1-4842-3931-5_8

dataset could be of garments trending on social media or other channels. The use of real-time data and generative models for design purposes could give companies like Amazon an advantage in understanding demand for garments before producing them. If you're interested in the **data-mining** aspect of this concept, read on to Chapter 9.

There are limitations to the current proposal for a computer-based fashion designer. Generative models can create images of garments, which is useful for providing an educated jumping-off point when it comes to trend research. However, any fashion designer reading this book knows that the reality of their job only begins with that image. An image of a garment is not a design, it is not a tech pack, and it is not a garment. These models do not have an inherent understanding of the real world; they are only able to identify patterns in data they are given.

GANs do, however, hold other potential for applications related to fashion. From graphics generation to automatic mapping of 2D images of garments onto images of people, the use cases are just beginning to be explored.

Artificial Creativity

Aside from Amazon's work in the area, other research has been exploring the possibility of generating fashion-based images as well. In their research paper, "DesIGN: Design Inspiration from Generative Networks," Othman Sbai et al. describe a method for generating garments by combining garment silhouettes as a mask and then transferring pattern and texture to that garment mask using a GAN. The goal of this research was not to propose the automation of fashion design, but to generate an inspirational machine assistant. Some of the most successful results from this research are shown in Figure 8-1.

Figure 8-1. *Images of garments generated by Sbai et al. from Facebook AI Research, Ecole des Ponts, and NYU*

The paper touches on some much larger philosophical topics in the field of AI as well. Creativity is a topic of controversy among those generating machine learning models that have creative motives. For many researchers, the idea that a machine can create original works of art and therefore embody a characteristic that is innately human is an ultimate achievement of artificial intelligence. For other researchers, it seems likely that creativity will never be achieved by a machine, but machines can provide excellent tools to the humans who are creating.

To date, in examples rendered, machines can only mimic creative works that they are fed. These include machine learning work that outputs images imitating great painters, the writing of pop songs, and doodling sketches. The reality of this work, as incredible as it is, is that it fails to meet our expectations when it comes labeling it as "art."

Mapping Garments onto Images of People

Think about all the different types of images that are created to sell a single garment style. Technical drawings, line plans, garment photos, garment-on-model photos, lifestyle photos, and blogger photos are all part of the strategy to get the idea of a garment into the hands of a consumer. Generative models provide one possible solution to reducing the cost of image generation in the fashion industry.

In "The Conditional Analogy GAN: Swapping Fashion Articles on People Images," Zalando researchers Nikolay Jetchev and Urs Bergmann propose a method for transforming 2D garment images onto images of people. An example of this work can be seen in Figure 8-2.

Figure 8-2. *A garment image (far right) being transformed onto a photograph of a person. The original source images are on the left.*

While the resulting images are still low fidelity and low resolution, this research shows promise for automating the process of photographing garment-on-model images. Photographing a season's garments can cost brands thousands of dollars per photoshoot. GANs have also been used for other types of photo editing, such as retouching.

Turning Sketches into Color Images

Another area of research using GANs is **image-to-image translations**. This includes converting a simple black-and-white sketch into a color image, a process usually described as edges to photo, as seen in Figure 8-3.

Figure 8-3. *An image of a black-and-white sketch on the left and an output photo-like image on the right generated by **pix2pix***

This kind of image-to-image translation is done using a conditional GAN. Less of the setup needs to be hand-engineered than was previously, making the model more accessible to use by a wider audience.

How Generative Models Work

On the most basic level, a generative model refers to a method of computer-driven generation of images, video, and music, for example. This is different from other types of machine learning because the output is a sort of recombinant variation of the training data.

The high-level concept is shown in Figure 8-4. A generative model takes in a series of input images, and outputs images that are similar but completely machine generated.

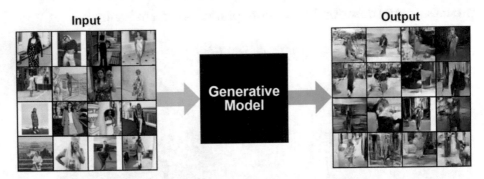

Figure 8-4. *A high-level view of what a generative model does. It takes in an input dataset and generates images that are "fake" but look similar to the original data.*

Generative adversarial networks (GANs), the most popular generative model and the method used by Amazon, are a subset of generative models. Put simply, a GAN is a series of two unsupervised dueling neural networks. In a GAN, one of the neural networks generates images based on patterns it recognizes in the input dataset, and the other neural network classifies those images as real or fake. The information is passed back through the generative network in order for the model to improve.

There are other types of generative models, including **variational autoencoders (VAEs)** and **autoregressive models**. A VAE relies on probabilistic modeling to generate outputs, but often the outputs of a VAE tend to be replications of the original dataset rather than creating something unique. Autoregressive models have remained probably the lesser explored of these generative models but are starting to become more popular.

Limitations

In a machine age, dressmaking is one of the last refuges of the human, the personal, the inimitable.

—Christian Dior, fashion designer

There is little reason to believe that a generative model will be able to write a tech pack anytime in the near future, for instance. It's also important to keep in mind that this is currently a topic being explored in the research community, and though the field is moving quickly, generative models are not quite ready for commercialization.

The AI fashion designer does touch upon some larger issues that many industries are calling into question. What is the role of human labor in this new era, as machines are taking over so many of the tasks that we are currently responsible for? What are the jobs we will hold as this becomes more true? Will the fashion designer be a profession of the past? What about other jobs in the fashion industry? This subject will be explored and speculated on in greater detail in Chapter 12.

Why GANs?

GANs show promise to AI researchers in part because of their ability to accomplish complex tasks using unsupervised techniques. GANs are also compelling because they provide potential solutions for complex automation problems in areas such as these:

- Photography

  - *Image in-painting and retouching*: Completing images with missing patches or even potentially retouching

  - *Increasing image resolution*: Converting images from low resolution to high resolution by filling in information

- Design
  - *Aggregating trends*: Generated by a target audience into a visual summary

  - *Style transfer*: Applying the style of a particular aesthetic quality

  - *Medium translation*: Taking images from a sketch-based medium to a photograph, and vice versa

- Storyboarding
  - *Text-to-image generation*: Eventually, trained GANs could be good enough to deliver images from a text-based input.

Admittedly, some of these applications are speculative, and this list is not comprehensive. The capabilities of technology like GANs and all of AI start and stop at what humans invest in building. Progress will require a motivated individual or team to pursue and accomplish many of these tasks. Without those people, it could remain dormant.

Implementation Example: AI Fashion Blogger

GANs are, at least currently, best suited for use cases that require images as the output. In the case of fashion design, this is challenging, because a garment is the final output rather than an image.

There are other avenues to apply GANs to the fashion industry; for example, the fashion blogger. The images in Figure 8-5 are output from a GAN I trained on fashion blogger data, images I scraped from blogger Instagram accounts.

Figure 8-5. *Examples of 64px images output from DCGAN trained on fashion blogger data*

Every image in the fashion industry has more to it than meets the eye. Fashion bloggers make up an industry of their own. Fashion blogger images could not exist without a complex network created by the people who made them.

On the flip side, fashion blogging wasn't an industry before the proliferation of two technologies: the smart mobile device and social media. It will be interesting to see the resulting outcomes from the introduction of AI to this ecosystem as a creator and how it will reshape the economy behind blogging.

How It Works

The easiest way to understand how a GAN works is through a description of a simple example. Figure 8-6 shows a dataset of 200 images. These images might be images from fashion bloggers' Instagram accounts. That data is used to train the first neural network, a generative neural network, which creates similar-looking images from scratch. The output dataset is a collection of 200 images that were created by the generative neural network (G). These are not real images of bloggers, but were created by the neural network to look like them. Once that dataset has been created, the second neural network takes the data as an input and returns the probability that the image is from the initial dataset. In other words, the second model classifies the image as real or fake. The goal of the GAN is to generate images that are convincing enough that the second neural network believes the generated images are examples from the real world rather than images that were created by the first neural network.

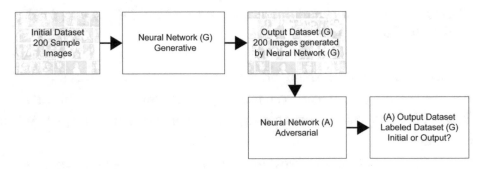

Figure 8-6. *A diagram of the example GAN*

Training GANs

Because GANs are made up of two neural networks, training a GAN is similar to training any other neural network. Refer to Chapter 4 for more information about how neural networks are trained.

Training GANs is still not a widely understood process. This GAN fashion blogger example was trained using a **PyTorch** implementation of a GAN called **deep convolution generative adversarial networks** (**DCGANs**). The first thing to do to train a GAN is to find or collect a large, high-quality dataset. Figure 8-7 shows the learning process using this method.

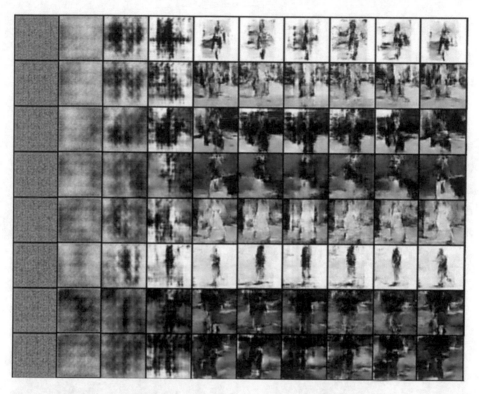

Figure 8-7. *A series of images that illustrate the learning process, from left to right. As the GAN trains, in every epoch it becomes better and better at generating realistic fashion blogger images.*

Each time the GAN iterates through an epoch, it learns how to improve the images, leading the discriminator network to believe that the images are real. An **epoch** is an iteration through the entire dataset. This terminology may be seen as less necessary or even potentially unnecessary in examples with large datasets, but in the case of this DCGAN, crucial to getting quality results.

Improving Results

The most straightforward variable to control while training these networks is the size and quality of the dataset. Expanding the dataset will improve the quality of images output by the network for a variety of reasons.

The more images there are, the less likely you are to run into issues of **overfitting**. Overfitting, in this case, occurs a when the data outputs too closely fit the inputs; sometimes this is caused by having too few images in the dataset. Hence, the network reproduces the same imagery in all of its output images. Figure 8-8 shows an example of overfitting; the same face appears over and over again in the generated images.

Figure 8-8. *An example of overfitting, expressed through the repetition of the same face over and over again*

For context, a researcher would consider CIFAR-10, a dataset of images of 80 million images, to be a large dataset. In the case of the fashion blogger, I was using a dataset of around 3,000 images to start.

The Future of GANs

This chapter just skims the surface of what can be done using generative models. While many of the image examples appear fairly low-resolution, some networks already can output images at higher resolutions and with more lifelike results. A couple of examples of networks capable of this are **StackGANs** and **progressive growing GANs** (**PGGANs**); sample PGGAN outputs are shown in Figure 8-9.

Figure 8-9. *Output images generated by PGGAN, released by Nvidia in late 2017. The model was trained on celebrity headshots.*

Today, these images are expensive to produce in terms of compute power and researcher hours, but generative models have quickly become a topic of interest in the machine learning community. Their use is growing and expanding to solve problems in new industries.

Summary

While today the promise of an AI fashion designer might be more hype than production ready, generative models have potential for a wide variety of applications in the fashion industry. From graphics generation and tools for creativity by Sbai et al. to automatic mapping of 2D images of garments onto images of people by Jetchev and Bergmann, the use cases in fashion are just beginning to be explored.

Generative models are still a topic of research but have gained traction quickly. For a deeper dive on generative models, I recommend the following resources. You can find even more in the annotated bibliography at the end of the book.

- "Generative Models" by Andre J. Karpathy et al., (OpenAI Blog, Nov. 28, 2017): `blog.openai.com/generative-models/`

- *Deep Learning* by Ian Goodfellow et al. (MIT Press, 2016): `www.deeplearningbook.org`

- "Photo Editing with Generative Adversarial Networks (Part 1)" by Greg Heinrich (Nvidia Developer Blog, April 20, 2017): `https://devblogs.nvidia.com/photo-editing-generative-adversarial-networks-1/`

Terminology from This Chapter

Autoregressive models—Train a neural network based on the conditional distribution of each pixel, given the previous pixel. This method is similar to methods used in NLP, but iterate over pixels instead of characters.

Data mining—A process of analyzing a large dataset in order to uncover patterns that might be prevalent. More on this topic in Chapter 9.

Deep convolution generative adversarial networks (DCGANs)—A type of GAN first described by Alec Radford et al. in 2016. Since then, many implementations of this neural network system have surfaced from the machine learning community.

Epoch—An iteration through a dataset. Iterating through a dataset more times, especially for small datasets, can help increase the accuracy of the outputs in GANs.

Generative adversarial network (GAN)—A type of generative model in which two neural networks are dueling, resulting in a higher-accuracy output in an unsupervised learning process.

Generative model—Often refers to a model that generates images, though generative models can also be used to create other types of data.

Image-to-image translation—Similar to a language translation in that it refers to a process of taking one style of images and converting it to another style.

Overfitting—Occurs when the data outputs too closely fit the inputs. In the abstract, it means that the underlaying algorithm is tailored to fit a specific set of characteristics and has less-broad application.

Pix2pix—Shorthand for a network architecture release by Isola et al. that allows users to do general image-to-image translations using a conditional GAN architecture.

Progressive growing GANs (PGGANs)—A GAN architecture concept that came out of Nvidia in late 2017. The outputs are 1024 × 1024px images that are indistinguishable from real images; their example was trained on celebrity headshots.

PyTorch—A Python (general-purpose programming language) framework for machine learning. In general, frameworks help software engineers to move faster in developing and refining machine learning models.

StackGAN—A method of using a system of multiple stacked GANs in order to generate higher-quality outputs, like higher-resolution images.

Variational autoencoders (VAEs)—A type of generative model made up of an encoder, a decoder, and a loss function. They work similarly to GANs in that they are also made up of a set of neural networks, measuring loss over the network generation and improving the model continually.

CHAPTER 9

Data Mining and Trend Forecasting

Fashions fade; style is eternal.

—Yves Saint Laurent, fashion designer

Trend forecasting has always been an elusive industry, with large brands paying hefty fees for consultancies to give them advice about the future. This forecasting toggles between art and science—with investment in advice ranging from groups like K-HOLE, a trend forecasting consultancy that started as an art collective making commentary about the corporate world, to researchers at Cornell University, who have taken to social media data to study fashion's anthropology around the world.

Today, social media platforms provide a place where information is aggregated, leaving a treasure trove for data mining. This data can be used to unlock the scientific piece of the puzzle, at least, when it comes to understanding and forecasting the cultural trends that drive the fashion industry.

Trend Forecasting

Trendy is the last stage before tacky.

—Karl Lagerfeld, fashion designer

© Leanne Luce 2019
L. Luce, *Artificial Intelligence for Fashion*,
https://doi.org/10.1007/978-1-4842-3931-5_9

The traditional methods of trend forecasting, including relying on sales data and trend reports, are often how brands gain an understanding of customer desires and behaviors. In today's age of digital personalization and social media, consumer demands are changing their purchasing behavior. Trends are bubbling up from culture rather than trickling down from a few key individuals. This change has made it difficult for marketers and designers to keep up with fleeting customer desires.

In the fashion industry, a season's design process might start with a trend-forecasting agency, like New York–based WGSN. The forecasted trends end up in the hands of dozens, if not hundreds, of designers in the industry. Color trends are propagated to textile designers and manufacturers, and the predictions can drive the supply more than the other way around.

In the meantime, consumers might already be expressing their perspectives relating to these trends in a public forum such as Twitter, Instagram, or Facebook. They might already be expressing boredom with a color that designers have chosen to be dominant in next season's palette.

Social Media

I have a great support network—my family, my modeling agency Storm, and people I work with in the fashion industry. And, of course, there are all my followers on Twitter who stop me from feeling lonely; I love them all. They keep me grounded.

—Cara Delevingne, model and actress

In 2017, researchers at Cornell University released a paper titled "StreetStyle: Exploring Worldwide Clothing Styles from Millions of Photos." The main hypothesis was that by referring to millions of images uploaded by users on social media, they could index local and global trends in what the users were wearing. Their approach took an anthropological angle of dissecting differences in garments through time, place, and style. Some of the findings of the paper might be too broad to create actionable trend insights for fashion brands, but the idea and many of the methods used in the paper provide a way for brands to use data from Instagram and other social media platforms to understand what their customers are wearing in real time.

Cornell wasn't the only university to take a look at social media and fashion in 2017. In fact, there has been a major uptick in fashion-based research in these fields. Doris Jung-Lin Lee et al. at the University of Illinois and a handful of others have presented research in this area as well.

Why has fashion become such a topic of interest? In part, the rising influence of social media on this industry means also a rising dependence on the technological platforms that make up the social media industry.

In many ways, social media has provided a way for more people to have a voice in an industry that was once heavily controlled by a few powerful gatekeepers. In this fragmented environment, however, the rise of influencer marketing has given power to a larger group of individuals who operate outside the confines of the fashion industry. These influencers represent niche groups of people within a community. Influencers' accounts act as central points of discourse about styles, colors, and other trending fashion details.

Social Media Mining

Social media platforms act as a source for data about how consumers feel about products available and products they've purchased. With that comes public discourse about, quite literally, everything. For the fashion industry, this data comes in the form of text, images, and video that can be used to predict trends. One of the most confusing parts about this process is knowing which data is valuable and how to get it.

There is probably no one who knows a brand's customer base better than that brand. To figure out which data is valuable, there's no single approach that will be right for everyone. What's useful about social media is that accounts are part of a network. We can learn about an account by the accounts that they follow and the accounts that follow them. In some cases, this is an effective way of finding the people who are most representative of the trends affecting your customers' behavior.

Figuring out where to look is the first step to finding valuable data. In "Identifying Fashion Accounts in Social Networks," Doris Jung-Lin Lee et al. use a snowball approach like the one described previously to find fashion accounts on Twitter. In the paper, they mention two techniques for getting initial access to data through Twitter: using Twitter's API and scraping tweets from account pages.

What Is Data Mining?

Data mining, introduced in Chapter 1, is the process of uncovering patterns in large datasets. There is a lot of overlap between the methods used in data mining and those used in predictive analytics, but they aren't analogous terms. Data mining methods and tools can provide building blocks for predictive analytics, but predictive analytics can extend further to other processes outside data mining.

Think of data mining as the step of discovering raw materials for a project. These are the components you'll need to build something. The process of data mining allows you to go through your collected data and algorithmically select only the data that is useful for the application you're working on.

To give an analogy, let's say you've collected a warehouse of fabrics, but they are stored in random order. You need to select the right fabric to create a lightweight blouse. You walk through the warehouse of randomly stored fabrics and have to determine, fabric by fabric, the features of each. Then you go through each to decide which you want to use. This process would take a really long time.

Instead, you can use data mining at this stage to select only the fabrics relevant to what you're trying to achieve. Maybe you filter it down by searching for lightweight fabrics, then fabrics with a satin weave. You could also find fabrics that have been used in blouses before or ones that have properties similar to other fabrics that have been used in blouses before.

APIs

How do you actually get the data you need? Most social media platforms offer public **application programming interfaces (APIs)**. APIs were briefly mentioned in Chapter 2 in relation to natural language processing. An API provides the building blocks of a program to the programmers using it. It can be used as the communication layer between applications.

In a common software web architecture, you might want to separate your **front-end** interface, the visual part that your customers interact with in a browser, from your **back-end** database and applications that contain your business logic. For these two separate areas to interact, you can create an API that acts as a bridge between the two. The API will have a bunch of **endpoints** that serve the specific purpose of allowing front-end user-interface components to interact with the data in the back-end database. A map of that kind of system might look something like Figure 9-1.

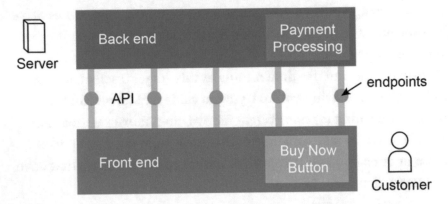

Figure 9-1. *A simplified diagram of a generic web application architecture*

Your customer sees the front-end interface hosted on a web site. They can do things like click the Buy Now button, for example, which would trigger payment-processing scripts in the back end.

Your back end might talk to other companies' APIs that are related to services you use. For example, Stripe is a common payment-processing service. In Figure 9-1, the payment processing app in the back end would connect with Stripe's API to carry out the task required. This is appealing because by revealing these endpoints, you can give access to programs running on your back end, allowing other services and platforms to be built on top of your API.

APIs can also do things like provide access to a database of content, like tweets. In the example of social media, companies such as Twitter have APIs where you can access tweet data in a form that's easy to collect and analyze programmatically. That means for a given topic or hashtag, you can go through public tweets, and using NLP techniques, analyze the general sentiment around a marketing campaign without ever having a human read a single tweet. There are, in fact, many examples of scripts people have written to do exactly this available online.

Web Scraping

For some applications, web scraping might provide an alternative for sites that don't provide a public API or for platforms that have limitations in their API.

Web scraping is a method of extracting large amounts of data from a web site programmatically, without using an API. The data can vary in format and subject, and include words, tables, hashtags, numbers, comments, images, and more. Web scraping is often an essential component in creating large datasets that can be further analyzed using machine learning and other artificial intelligence techniques. Web scraping is typically less reliable and more difficult to implement than programmatic access over an API. Web scrapers depend on the structure of the web site staying the same. If the site changes even in a small way, that change might break the scraper.

As a disclaimer, many companies do not allow you to scrape their web sites without their consent. Unmindfully scraping can also lead to site-reliability issues: if you try to get too much data at once, you might break the web site for other users. Scraping can get you into legal trouble if a company's Terms of Service forbids it. You might consider consulting a lawyer if you're planning ambitious web scraping projects.

BETTY & RUTH TREND RESEARCH

Quality data is really important to all of the technology we have implemented at Betty & Ruth. We frequently use web scraping to help acquire the best possible datasets for our trend research.

Let's say we're looking to find images from Instagram and we want to find accounts that relate to the target demographic of our brand. For Betty & Ruth, our goal is to discover trends and predict what our customers will buy next season.

Our strategy might be to look to acquire data from particular influencers whose audiences correlate strongly with our customers. We could identify accounts that use certain phrases in their description or hashtags, like these: fashion blogger, style blogger, ootd, sheco, minimalist fashion. Then, we might refine that group of influencers even further by searching for particular locations, follower count, or filter by accounts that mention certain brands. There will likely be several stages to narrow the data to exactly what we're looking for, even before we get to the images themselves.

After that, we can use those specific accounts and scrape the data from their pages or access Instagram's API. If we're looking to collect a lot of data, we can write code to do this. If we don't need a lot, there are browser extensions that we can use to dump images from a web page into a folder on our computers.

Web Crawlers

Web crawlers, or **site crawlers** (mentioned in Chapter 2), provide an automated method of extracting information from web sites in a systematic way. In the event that your project requires data that will be updated daily, a web crawler might be the most effective means of collecting this data.

Web crawlers are an essential tool on the Internet. Google and other search engines use web crawlers to give constantly updated information on their search pages. You'll often hear people refer to web crawlers as a spider, because they "crawl" through the Web and examine sites.

The way data mining and web crawlers work together is that web crawlers can be used as a first step in data mining, to collect the raw data that is desired.

Data Warehousing

Where does all this data get stored after it's collected? The answer is most likely that it's being stored in a **data warehouse**. A data warehouse is a software-based central place where a company's data is stored. It can contain one or more databases and is designed to store large amounts of data. Data warehouses are optimized for reporting and analysis.

The hardware that the data warehouse lives on is called a **server**. Servers are computers designed to store data and process requests. These servers are usually housed in a **server farm**, also called a **data center**. Figure 9-2 illustrates the relationships between these concepts.

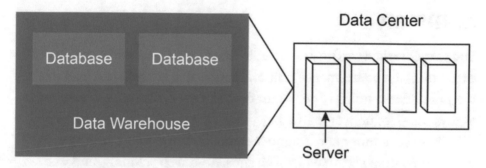

Figure 9-2. *Relationships between a database, data warehouse, server, and data center (or server farm). On the left are the components that exist as software components. On the right is the physical hardware that holds those software components.*

For quick projects and experimentation, you do not need to set up a data warehouse. Data warehouses are an optimization used when you're looking to scale solutions that have already been proven to work.

For production, data will be fed through a data pipeline. A data pipeline is a series of programs that transform data from one state to another. In this example, the process transforms raw data into data that analysts can use to create trend reports. The data starts at web sites, third-party sites, and APIs, and then is moved to the server or program to be processed. From there, it is stored in a data warehouse, where analysts can access it to create trend reports. This process can be seen in Figure 9-3.

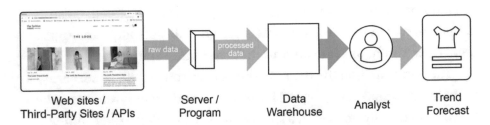

| Web sites / Third-Party Sites / APIs | Server / Program | Data Warehouse | Analyst | Trend Forecast |

Figure 9-3. *A simplification of a data pipeline*

Summary

Social media data can be used to gain insights about specific groups of users on the platform. For fashion brands, this analysis can lead to a deeper understanding of customer desires, especially through sentiment analysis and influencer marketing.

There are a number of methods available to use for extracting data, particularly from social media web sites. APIs, web scraping, and web crawling can be applied, depending on the requirements of the project and the platform you're looking to gather information from. Each of these methods of data collection addresses a different need. While APIs give easy access to structured data on some platforms, they may have limitations on the data and how you use it. Web scraping can give more-flexible access than APIs, but it might have engineering and legal limitations. Web crawlers provide repeat access.

With all of this data being collected, data warehousing is a common approach to storing it. This method is optimized for analysis and reporting as opposed to other types of storage, which might be better at simply retrieving information or running scripts.

Terminology from This Chapter

Application programming interface (API)—A communication layer of a software architecture.

Back end—The part of a software or web architecture that typically contains data and business logic. The back end might run simple scripts to complete tasks or return information to the front end. It might also run scripts that connect with external services such as payment processors.

Data center—A physical location where servers are maintained and run. See also *server farm*.

Data warehouse—A place where data resources are stored, usually in the form of databases. This type of storage structure is optimized for analysis and reporting.

Endpoint—An access point of an API that allows programmers to write code that uses back-end scripts and services.

Front end—The graphical user interface of a web application or other software applications.

Server—A computer or program typically used to provide functionality like information or services to other devices, called clients.

Server farm—A physical location where servers are maintained and run at scale. See also *data center*.

Site crawler—Scripts that are run to collect data from web sites in a systematic way. They're useful for projects that require data that is being constantly updated. See also *Web crawlers*.

Web crawlers—See *site crawlers*.

Web scraping—A generic term that refers to extracting data from a web site.

PART V

Supply Chain

CHAPTER 10

Deep Learning and Demand Forecasting

Machine learning drives our algorithms for demand forecasting, product search ranking, product and deals recommendations, merchandising placements, fraud detection, translations, and much more.

—Jeff Bezos, CEO, Amazon

Demand forecasting is a branch of predictive analytics that focuses on gaining an understanding of consumer demand for goods and services. If demand can be understood, brands can control their inventory to avoid overstocking and understocking products. While there is no perfect forecasting model, using demand forecasting as a tool can help fashion businesses better prepare for upcoming seasons.

For fashion brands, estimating the amount of inventory to manufacture can be a complex game of combining historical data, applying intuition, and forecasting fashion trends. Placing too large an order of a particular garment style can lead to inefficiencies for the brand, from decreasing margins when items are discounted on the retail floor, to complete loss if the products don't move. For these predictions, investing in demand forecasting strategies could be worth billions of dollars every year for the fashion industry. The waste created by overproduction is both a financial and an environmental disaster.

© Leanne Luce 2019
L. Luce, *Artificial Intelligence for Fashion*,
https://doi.org/10.1007/978-1-4842-3931-5_10

While demand forecasting remains a challenge, developments in machine learning have provided dramatic improvements. In this chapter, we'll explore deep learning in demand forecasting for fashion.

What Is Demand Forecasting?

Forecasting as a general topic in fashion can address needs such as capacity planning, stock keeping, and pricing strategy. In simplest terms, demand forecasting estimates future sales. Forecasting can encompass long-term planning (for example, how much revenue will we generate this year, and how should we grow our workforce next year) as well as short-term planning (for example, how many shirts of this style will be purchased this year). In this chapter, we will discuss the latter, though many of these techniques can be applied to both.

Capacity planning and stock keeping usually fluctuate over time. The forecasting techniques to predict these future demands rely on **time-series data**, data that is successive over a period of time. This type of prediction is referred to as **time-series forecasting**.

Forecasting Methods

Demand forecasting is an entire field on its own. There are many methods for forecasting demand, ranging from more qualitative to quantitative approaches. Examples of methods used are averages, time-series analysis, surveys, as well as machine learning approaches including deep neural networks. Depending on the data, some methods work better than others.

Fashion's Challenges in Forecasting

Across retail sectors, demand forecasting has become a critical tool. For grocers whose products may expire over a few days, predicting demand can radically change profit margins. For fashion, the idea is not much different, though in some ways it is more complex. There is a lack of historical data to rely on, because new items are being constantly introduced, and seasonal trends add an extra layer of unpredictability. The challenges to overcome in fashion include overproduction, expiration, fast and short seasons, and unpredictable consumer behavior.

Overproduction

As the pain point of overproduction has become increasingly detrimental to apparel brands, the entire industry has begun to shift to accommodate. One of these changes is a movement toward vertical integration, which has led to faster response times across an organization. Some experts have suggested that the best solution is to be building flexible manufacturing infrastructure that allows the industry to respond to changing demands in real time. While flexible manufacturing does provide a solution to the same problem, it's not necessarily a substitute. Improved forecasting can provide shorter-term relief and reduce waste and spending.

Fast and Short Seasons

Understanding demand is so critical in the fashion industry partly because the product life cycle is short, the selling season is short, and the replenishment lead times are long. For major retailers like Zara, there are around 20 seasons in a year, leaving only a couple of weeks to sell the styles in a season.

Although fashion goods do not expire physically, the desire for those goods does. The timelines have shortened as the movement of goods in fashion has increasingly taken on the patterns of a consumable.

In some ways, this is quenching a thirst for creativity and self-expression. In other ways, it may be illuminating a toxic aspect of our culture. The more we consume in fashion, the more waste we generate.

Consumer Behavior

Consumer behavior in this space is emotionally driven and often highly impulsive. Purchasing patterns are also highly impacted by external factors. They're volatile in that even the weather and the news can affect sales dramatically. Aside from all of these challenges, the product itself has a wide range of variety in style, color, surface treatment, print, and so many other variables.

Nonforecasting Solutions

Without the ability to accurately forecast demand of individual fashion items, brands and retailers have turned to measures to create demand. One of the most straightforward ways to create or control demand is through price.

Price Prediction

One common variable to optimize in sell-through of a product is price. If you're selling through a product every time without optimizing the price, there might be an opportunity to charge more for that item. On the flip side, if an item isn't selling, decreasing the price might be a simple change to help increases sales of that item. Markdowns and promotions are ever-popular techniques for addressing inventory management in the fashion industry.

Historically, pricing strategy has relied on a mix of markup cost, competition pricing, and the merchant's judgement or intuition about what a customer will pay for the item. This strategy has its shortcomings, as it relies on methods that are often immeasurable.

We're mentioning price here as a nonforecasting strategy to manipulate demand, but the relationship can also go the other way around. Forecasting can be used to determine the optimal price for a product by interpreting related information including competitor data, historical data, or sales of similar items.

Deep Learning

The analogy to deep learning is that the rocket engine is the deep learning model, and the fuel is the huge amounts of data we can feed to these algorithms.

—Andrew Ng, general partner, AI Fund

How can deep learning help resolve some of the complex challenges in demand forecasting for the fashion industry? Deep learning has been used to handle complex problems with many variables that are difficult to model manually. These techniques work best in a data-rich environment, or in use cases in which there is a lot of data.

Deep neural networks have the potential to generate more-accurate predictions than a linear model because of compatibility when it comes to the complexity of these problems. One of the major downsides to a deep learning approach is that it's difficult to interpret how the algorithm arrived at its results, which are crucial for business decisions.

What Is Deep Learning?

Deep learning is a specialized subset of machine learning. Deep learning usually refers to large neural networks trained on large datasets. Deep learning is being used for a vast array of applications, not only inside the fashion industry and retail, but in medicine, self-driving cars, and beyond. While deep learning is often recognized for successes in image generation, like the GANs described in Chapter 8, it can also be used for other tasks.

In fact, several of the techniques discussed in this book can be considered under the umbrella of deep learning, such as deep neural networks, convolutional neural networks, and recurrent neural networks.

The distinction between basic machine learning and deep learning is difficult to define. One explanation is that basic machine learning tends to be more task specific, and deep learning more learning based. Deep learning encompasses neural networks that have more hidden layers.

Traditional methods often rely on **linear regression** models for demand forecasting and have had less success in fashion because of the chaotic nature of the industry. While these might give some high-level estimation to demand, they don't give enough resolution to make predictions about what will happen in the future.

In order to get a high-resolution forecast, the predictions need to fit the curve that demand does. Much of machine learning is focused on **curve fitting**, with the goal to more accurately track reality as represented by a graph. Figure 10-1 shows how the real world, a linear model, and a machine learning model might look.

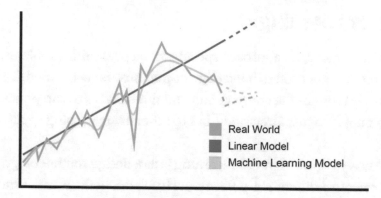

Figure 10-1. *In a machine learning model, complex curve fitting can track closer to the real world. In a linear regression model, the average turns out to be an oversimplification.*

Deep Learning for Demand Forecasting

Neural networks provide greater flexibility in demand forecasting because they are nonlinear models that can take in a lot of variables and then output simple predictions. However, their benefits are greatly affected by the training data. Without a lot of high-quality data, it can be difficult for these models to perform. This is actually a major inefficiency of using deep neural networks for the application of demand forecasting; they are not very data efficient.

Long short-term memory (LSTM) is one of the most talked about deep learning models used for time-series forecasting. LSTMs are a type of recurrent neural network with unique memory-like behavior that allows them to learn patterns better.

Techniques for Smaller Datasets

While data is an important aspect of all machine learning, companies with small amounts of data can still apply deep learning through **transfer learning** and other forecasting models that are specialized for smaller datasets.

Transfer Learning

In transfer learning, an approach specific to deep learning problems, pre-trained models are used to bootstrap another more specific model. This significantly reduces the computation and time it takes to carry out tasks. It doesn't apply to just forecasting, but to other deep learning applications as well.

One way to think of transfer learning is as a design methodology. It doesn't refer to a difference in the model itself, but in the way it is trained and run. Rather than starting from nothing, you can start from a model that is trained on data that was used for a similar task. For example, if I trained an image classifier to identify whether an image contained a sneaker, I could use that pretrained model to recognize other items (such as boots, maybe). The differences in design methodology are expressed in Figure 10-2.

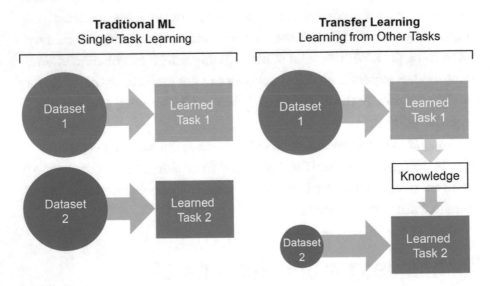

Figure 10-2. *Comparing transfer learning to traditional methods of machine learning*

Transfer learning isn't just a solution for small datasets, but can be used as a way to make the most of older data that might have less relevance in making predictions. What a brand was selling in 1990 might be irrelevant to them today, but the data can still be useful in training a forecasting model.

Other Forecasting Models

There are conflicting experiments and opinions about which models work best for forecasting applications. While deep learning methods such as LSTM appear to work well for applications with a lot of data, other models could outperform LSTM in certain use cases.

This chapter focuses on deep learning as a demand forecasting method in order to explain deep learning. However, there are other models for forecasting, including **autoregressive integrated moving average (ARIMA)** and **Prophet**.

ARIMA

Each of the components of ARIMA explains what the model does. Autoregression (AR) shows a changing variable, integrated (I) refers to taking the difference in the equations between current and previous values, and **moving average** (MA) incorporates a lagged average, such as a three-day average in which each point on a graph contains the average of the previous three days. Until recently, this was the state-of-the-art method for forecasting.

Prophet

In 2017, Facebook released its open source forecasting model, Prophet. In comparison to the data-hungry models presented by deep learning, Prophet is more data efficient. Prophet is also really useful for data that is seasonal.

Prophet represents an analyst-in-the-loop approach that can be thought of as human analysis augmented by machine tools. When you create a model, a trained analyst who understands the data can make it better. This is important because quality forecasting requires a highly specialized person, but there are very few of them. The Prophet model bridges the gap by getting the machine to do fairly good work, and having a less-trained analyst who can fill in the gaps. In this approach, the barrier to getting a good forecast has been lowered. This is also useful because it allows anomalies, such as holidays and other features that would be difficult to model, to be input by the analyst.

Each method has its trade-offs, Figure 10-3 shows four high-level approaches and their pros and cons.

Model	Pros	Cons
Linear Regression	· Easy to understand · Handles different components	· Sensitive to outliers · Strong assumptions
ARIMA	· Easy to understand · Fits historical data well · Forecasts unbiased	· Sensitive to outliers · Small forecast range
Prophet	· Easy to understand · Analyst in the loop · Data efficient · Fast	· Sensitive to compounding seasonality · Required data format
Deep Learning *Neural Networks* *LSTM* *Transfer Learning*	· Can take in many complex variables · Finds nonlinear patterns · Strong predictions · Easy to automate	· Difficult to understand · Requires a lot of data

Figure 10-3. *Pros and cons of the approaches discussed in this chapter*

While deep learning and other machine learning techniques are useful in the context of demand forecasting, they're just a tool. A highly skilled analyst in a particular industry has deep domain knowledge of a space that is often not captured in the data or that might go against the data at times. Rather than ignoring the knowledge of the analyst in these contexts and blindly following the machine's predictions, some models (for example, Prophet) keep the analyst in the loop.

Summary

Demand forecasting is a long-standing challenge, especially in fashion, which requires inventory and resource planning for the production of physical goods. Short seasons and irregular customer behavior make demand even more difficult to predict in this industry.

The use of deep learning models for demand forecasting is still a nascent field. Each year, new research is presented and improved upon. There is no perfect forecasting model, but these tools can be helpful predictors of demand even in volatile markets.

Demand forecasting is a complex and nuanced field. You can learn more from the following books and from references in the annotated bibliography at the end of this book:

- *Forecasting: Principles and Practice* by Rob J. Hyndman and George Athanasopoulos (OTexts, 2018)

- *Applied Predictive Modeling* by Max Kuhn and Kjell Johnson (Springer, 2016)

Thank you to Adam Bouhenguel, founder of Tesserai for providing expert advice in machine learning and demand forecasting as background for this chapter.

Terminology from This Chapter

Autoregressive integrated moving average (ARIMA)—A common approach to forecasting time-series data that does not involve machine learning.

Curve fitting—The method of finding an equation that most closely models the data. Compared to linear regression, curve fitting requires more-complicated formulas to represent the relationships between data points.

Deep learning—An area within machine learning usually referring to neural networks composed of many layers.

Demand forecasting—Part of predictive analytics that focuses on understanding and predicting future demand for goods and services using models trained on historical data.

Linear regression—A simple summarization tool in statistics in which a series of points on a map are averaged by plotting a line through them.

Long short-term memory (LSTM)—Neural networks composed of LSTM units. These units are made to include a mathematical representation of memory between units. Rather than a linear path through the network, an LSTM creates different paths based on input, output, and forget gates associated with the model. It gives the network a memory-like quality.

Moving average—Takes previous intervals and averages them in an attempt to more closely represent a trend over time.

Prophet—An open source forecasting tool released by Facebook in 2017. It's an additive regression model in which certain variables can be adjusted easily by an analyst to improve predictions.

Time-series data—Data collected successively over a set time interval.

Time-series forecasting—Predicting future events based on data that is collected over time.

Transfer learning—A methodology in which a neural network is trained on one dataset, and then the knowledge is stored and then applied to a different but similar problem.

CHAPTER 11

Robotics and Manufacturing

The topic of robots can be a fantastical or functional conversation, depending who you're speaking to. For a long time, science fiction has been dreaming of the ways in which robots might look like us, talk like us, think like us, and take over the world. In reality, most of the machines being built in the field of robotics have nothing to do with the portrayals on television. The most useful robots in industry today are not humanoid, aren't **bipedal** (walking on two legs), don't speak, and don't think like humans.

Robots in Popular Culture

We're fascinated with robots because they are reflections of ourselves.

—Ken Goldberg, professor, UC Berkeley

Robots in popular culture have given consumers the image of a humanoid robot that looks as beautiful as Scarlett Johansson or is an assassin, as in *The Terminator*. We should challenge these portrayals for a number of reasons: they're both inaccurate and have a negative social impact.

© Leanne Luce 2019
L. Luce, *Artificial Intelligence for Fashion*,
https://doi.org/10.1007/978-1-4842-3931-5_11

By depicting robots this way, we do not accurately represent what robots are capable of. Media portrayals ignite fear instead of sparking curiosity. In medicine, the Da Vinci robot, shown in Figure 11-1, is capable of completing operations with higher precision than a human alone. In space, the Opportunity rover, also shown in Figure 11-1, was built to survive just a few weeks but has been pinging back video to Earth for over 13 years. These robots look nothing like the replicants from *Blade Runner* but have led to progress in their respective fields without posing any threat to humanity.

On the flip side, building humanoid or anthropomorphized robots has proven to increase consumer trust in these machines. In certain contexts, like elderly care, this can be a positive goal to work toward if we can get past the **uncanny valley** of it all.

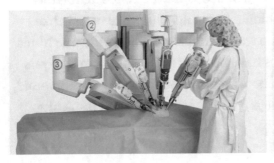

Figure 11-1. *The Da Vinci robot (left) and the Opportunity Robotic Rover (right). Images courtesy of Da Vinci and NASA/JPL, respectively.*

Robots and Women

I can love, I can speak, without somebody else operating me.
You gave me eyes, so now I see. I'm not your robot. I'm just me.

—Miley Cyrus, singer and songwriter, lyrics from *Robot*

The portrayals we create of robots as humans have ethical consequences for the professional and personal lives of many people. One of the most obvious examples of this impact is through gendered representations. I am not the first to call out that the patterned portrayal of women as sexy, nonthreatening, and passive robots in both Hollywood productions and as AI assistants is unethical.

We often assign AI assistants female names like Alexa, Siri, or Cortana. We see films with sexualized female robot characters like Rachel in the original *Blade Runner*, Ava from *Ex Machina* in Figure 11-2, or Samantha from *Her*. We even build female robots for entertainment or pleasure, like Sophia from Hanson Robotics in Figure 11-2, Project Aiko, EveR, Actroid, and so many more.

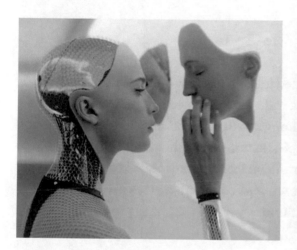 

Figure 11-2. Ava from the feature film, Ex Machina (left), and Sophia from Hanson Robotics (right). (Sophia image courtesy of ITU Pictures from Geneva, Switzerland.)

This is not to say that there are no male robots. HAL 900 from *2001: A Space Odyssey* is just one of many examples. The question is why the robots represented as female are sexualized servants rather than having any accomplishments of their own? Why is it that there is no masculine equivalent word for the term **fembot**?

You may disagree with this premise, but the point is simply that none of us are exempt from responsibility when bringing these ideas and machines into the world. As creators, we should think about what we're presenting and the impact that it has on the psychology of the people around us. There are already biases prevalent in our culture. Proliferating these ideas within devices that we live and work with is leaving an impression on their users. The creators are the individuals who are responsible for the impact these devices have.

It is rather humorous to reflect on humans being the only species on earth that could create a machine that looks like them and be fooled by it.

What Is a Robot?

The term *robot* has been applied loosely to many kinds of devices. There is a constant debate even among industry experts as to what truly defines a robot. For some, it is simple enough to describe a robot as a programmable machine that carries out complex actions. For others, a robot is really the physical embodiment of artificial intelligence that takes action in the physical world. The word *autonomous* comes up a lot as a descriptor. In short, there is no perfect definition.

Types of Robots

There are many ways to categorize robots. They can be organized by industry: medical, space, entertainment, and manufacturing, for example. They can also be organized by the way they move: by wheels, walking, flying, swimming, or stationary, not moving at all.

In manufacturing, the types of robots that we're usually referring to are **industrial robots**. These are robots that are specialized for functional tasks, like moving and assembling parts in a factory or warehouse. They're usually stationary and by means of the types of tasks they carry out are often mounted to the ground.

Industrial Robots

Industrial robots were originally built for a wide range of tasks, usually relating to rigid mechanical parts. The types of operations these machines typically perform are things like welding, assembly, inspection, packaging, and deburring. These relate to materials like plastics, metals, and cardboard, on a basic level in sheet or block form.

Articulated Robots

The most prevalent industrial robots are **articulated robots**. Usually this type of robot is a jointed arm with several **degrees of freedom**. Articulated robots can be used for very complex tasks. The body of these robots is referred to as a **manipulator arm**.

Choosing the correct robot for a particular task is important. Most robotic arms have a low **payload capacity**, which just means that they cannot lift things above a certain weight. In fashion, this is generally not an issue, as garments usually weigh under a couple of pounds. In other industries that deal with heavier materials like blocks of metal, payload capacity can be a major constraint.

There are typically five major components on a robot: a **control system**, sensors, **actuators**, **power supply**, and **end effectors**. Figure 11-3 shows a diagram of these components on an articulated robot. The control system and sensors are not shown in the image because these components can be mounted in various locations on the arm. The control system (also called controller) is often located in the base and sensors may be integrated throughout the arm depending on application.

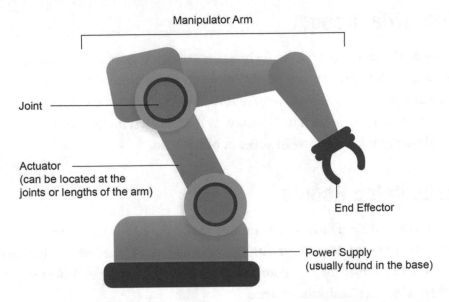

Figure 11-3. *A simplified diagram of a manipulator arm based on its major components*

Material manipulation is handled by the end effectors. An end effector is the "hand" of the robot arm. In robotics, choosing the right end effector for the task is akin to choosing the proper foot on a sewing machine.

End Effectors

End effectors often play a major role in determining whether a technology can handle a material. For instance, an end effector that will be used to pick up screws is a very different tool than one that would pick up grapes. In fashion, this is especially challenging due to the way fabric deforms when it is moved. Picking up a piece of fabric leads to an infinite number of possibilities in terms of the specific shape and number of creases the fabric might take on. Figure 11-4 shows different types of end effectors that might be used in a traditional robotic system like grippers and vacuum cups. Tools like screwdrivers, welding guns, drills, and spray guns are also common end effectors in manufacturing environments.

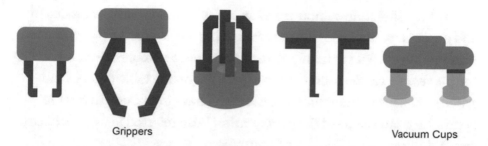

Grippers   Vacuum Cups

Figure 11-4. A variety of end effectors, such as grippers and vacuum cups, used for different applications in robotics

Sewing Robots

SoftWear Automation, an Atlanta-based robotics company focused on sewing, uses a combination of a table and manipulator arms with end effectors to move fabric both during and between sewing operations. In this system, the table itself is actuated with a series of vacuum-based components that move up to four layers of fabric, depending on the weight and pile. A close-up of the sewing head can be seen in Figure 11-5.

Figure 11-5. A close-up of the SoftWear Automation sewing head, which uses computer vision to guide the fabric through sewing

By implementing computer vision at the head of each sewing machine and at other points in the system, the sewing tables created by SoftWear Automation are able to manipulate and sew even complex curves. Keeping the pieces of a garment or other soft good flat on the table reduces the amount of deformation that the system will have to manage in sewing. Figure 11-6 shows one of the **gantry robot** (also referred to as a **cartesian robot**) tables used at SoftWear Automation.

Figure 11-6. *A gantry-based system for moving fabric pieces across the sewing table*

SoftWear Automation uses multiple techniques in a sewing system to completely automate the production of specific goods. Figure 11-7 shows both the gantry-based system for moving textiles across the table and a manipulator arm that is mounted to the gantry and can pivot the fabric to bind the edges during sewing.

Figure 11-7. *The SoftWear Automation sewing system for binding the edges of car mats*

The deformation of the materials is not the only challenge in robotic sewing. Sewing operations are incredibly high speed, at around 5,000 stitches per minute. To navigate these challenges, the hybrid systems built by companies such as SoftWear Automation are different from other types of robotic manufacturing.

Advantages of Robotics in Sewing

Palaniswamy Rajan, chairman and CEO of SoftWear Automation, has presented several advantages to robotic sewing: cost reduction, **reshoring** production, addressing globalization, environmental impact, and adding manufacturing flexibly to resolve issues like fit.

In the United States, most of our manufacturing has been outsourced to overseas vendors. Over time, the areas of the world that once provided high-quality and inexpensive sewing labor, like China, are growing into economies that look much like the United States itself. That means the wage delta is decreasing over time and so is the willingness to do tedious and taxing labor for low wages.

There are other inefficiencies in the textiles-to-garment industries. America is the third largest producer of cotton in the world. Yet, most of the goods that we consume made of cotton have been imported. Typically, we send our raw cotton goods overseas to be turned into textiles, sewn into garments, and then shipped back, because that is more cost-effective in terms of dollars. One of the unfortunate consequences is the environmental impact that has. While it may seem obvious that we could simply manufacture these goods in the United States and avoid the complicated logistics, manufacturing in America has decreased dramatically over the past 50 years, and we could neither compete on cost or volume with the current reliance on labor. Robotic automation of sewing tasks provides a way to decrease labor costs and increase throughput while manufacturing local to the source of raw materials, in the United States and other nations.

Each garment can be made one piece at a time using these robotic systems. In contrast, traditional manufacturing occurs in batches that have minimum quantities per size in order to meet cost efficiencies. By making each item one at a time, certain variations can occur without a change in the cost efficiency. Each can be a different size, and to the robot, it doesn't make a difference.

Designing for Robots

In this new era of robotic production in the soft goods space, we'll have to keep in mind new design principles when it comes to designing for robotic manufacturing. Over time, we will see an increase in the use of robots for apparel manufacturing.

It's also important to bear in mind that as a field, robotic sewing is still nascent. In terms of design, not every variation of every type of garments is possible just yet. For now, brands will have to keep things simple.

Automation and Robotics

Robotics and **automation** are not synonymous. Automation can occur in software alone, not requiring physical robotic hardware at all. Robotics can also be used to do things other than automate tasks, though they are commonly used for automation. Automation is primarily with the goal of having a machine to automatically complete tasks. Robotics generally has a more responsive approach and a goal to gain results within an imperfect world.

The distinction is usually that robots can be used to carry out a range of complex tasks, not just a single operation. They can also sense and react to real-world inputs.

On the flip side, industrial automation techniques may refer to a pocket-setting, which can only set pockets. If something should be set up incorrectly in those manufacturing environments, the machine will not respond. The autonomy of a robot to respond based on sensing the real world is part of what makes it a robot and distinct from other types of automation.

Questions of Responsible Automation

The conversation around robots replacing people in the workplace has seeded concern around how to introduce responsible automation. What will employers do to train the workers they are currently employing for new jobs rather than displacing them? Will robotic automation support a movement toward made-to-order manufacturing rather than mass manufacturing? There are many questions for a future fashion industry to answer.

Supply-Chain Robotics

Robots are being used beyond the factory floor in the fashion industry. Already, in warehouses around the world, robots have made their way into **picking and packing** procedures. From Tencent and Alibaba to Amazon, billions of dollars have been invested in building robotic automated infrastructure to manage the redundant task of finding inventory items and putting them into a box to ship to customers.

Kiva Systems is a well-known example of robotic automation in warehousing. They were acquired by Amazon in 2012 and renamed Amazon Robotics. Before the acquisition, they were being used by major corporations across the fashion industry, including Saks Fifth Avenue, Gilt Groupe, The Gap, and more. Today, there are a number of companies like Kiva Systems such as AutoStore, Dematic, and OPEX.

These robots are typically considered to be part of a school of thought that robots should be working alongside humans rather than replacing them. This idea is commonly referred to as **collaborative robotics**.

Historically, industrial robots have been large, metal robots. This made them dangerous to humans, because a human who stood in front of a robot could easily be hit unknowingly by a robot arm. Some collaborative robots, soft robotics, are made from soft materials to avoid this. Others may be equipped with better sensing mechanisms and are lighter weight and lower strength.

Lights-Out Manufacturing

The concept of collaborative robotics has helped companies more flexibly integrate robotics into their processes, allowing humans to complete the last-mile tasks. As robots have become more affordable than ever, cost reductions can be made by automating more and more processes in factories and warehouses. For some, removing the human from the processes altogether is the ultimate goal.

A popular conceptual goal in manufacturing is to work toward **lights-out manufacturing**. The idea is that the factory is fully autonomous and can operate at full clip even with the lights out. Some factories have even achieved this goal. There are efficiencies that also come along with building a factory like this. The location no longer needs to be based on the availability of a large-enough work force, but solely around trucking logistics to deliver that product from the factory location.

Summary

While pop culture has most people fantasizing over humanoid robots that talk, walk, and think like humans, the robots making the largest impact in the fashion industry have little in common with these portrayals. Industrial robots, cartesian robots, and collaborative robots make up a large portion.

There is a growing presence of robotics in the fashion industry, particularly in manufacturing and warehousing settings. As robots are developed to handle increasingly complex and sewing-specific tasks, the fashion industry will be confronted with new challenges of ethical automation, designing for robotic automation, and reshoring production.

Terminology from This Chapter

Actuators—The "movers" in a mechanical system. In the systems described in this chapter, they are the component that allows the machine to move, given a control signal. They include things like lights and speakers—anything that transforms a signal into something that happens in the real world.

Articulated robots—Can be industrial, legged, or others types of robots that have joints, often powered by electric motors, which give them a complex range of motion.

Automation—The use of equipment or machines to automatically complete processes.

Bipedal—To walk on two legs, usually in reference to animal locomotion.

Cartesian robot—A robot that is limited by x, y, and z motion. They're also sometimes referred to as *linear robots* or *gantry robots* because of their visual resemblance to gantry cranes. Many CNC machines and 3D printers follow this type of architecture. See also *gantry robot*.

Collaborative robotics—Robots that can safely work alongside humans. The safety required is achieved through tactics such as sensing, speed, and power and force limiting.

Control system—The brains of the robot. The control system is where sensing occurs and decisions about actuation are made in response.

Degrees of freedom—The number of directions a device or robot can move in. The range of motion can have a large impact on which tasks the machine can carry out. In a human body, the shoulder has three3 degrees of freedom: pitch (up and down), yaw (left and right), and roll (rotation).

End effectors—In a robotic arm, these are analogous to hands. They are responsible for much of the dexterity of articulated robots.

Fembot—A robot that looks like a female that exists solely or primarily as a sexualized character or toy.

Gantry robot—See *cartesian robot*.

Industrial robots—Come in multiple shapes and sizes, many of which include manipulator arms or gantries. They have been historically used in industries dealing with rigid materials (for example, metals and plastics) in applications including automotive, warehousing, and more.

Lights-out manufacturing—The idea of automating every process in a factory so that no human is needed to keep the system in motion.

Manipulator arm—The arm of an articulated robot. See also *articulated robot*.

Payload capacity—How much weight a machine can carry.

Picking and packing—A commonly used term for the process of locating items to ship in a warehouse and packing them into a box for customers as part of fulfillment.

Power supply—The source of electrical power that enables the operation of a machine. Typically, this device also provides current conversion to provide the correct voltage, current, and frequency to the machine.

Reshoring—Bringing domestic manufacturing back to a country. This is the opposite of offshoring, which is the process of manufacturing in countries overseas, where labor is cheaper.

SoftWear Automation—A corporation based in Atlanta, Georgia, focused on robotic automation for sewing soft goods. The company began as a research initiative at Georgia Tech. After seven years of research, the company spun out to begin providing manufacturing work-lines for sewn goods in home goods, footwear, apparel, and automotive.

Uncanny valley—A term coined to capture the unsettling nature of machines that look like humans. The idea is that robots that have a human likeness, but are apparently not human, can elicit a feeling of revulsion from humans. The hypothesis is that the more unmistakably human a robot looks, the greater the emotional response from humans. In order to get to that point, the likeness needs to pass through this threshold to overcome the creepiness.

Thank you to Palaniswamy Rajan, chairman and chief executive officer at SoftWear Automation for answering questions about sewing and robotics for this chapter.

PART VI

Future

CHAPTER 12

Democratization and Impacts of AI

A diversity of thought, perspective and culture is important in any field.

—Sarah Friar, chief financial officer at Square

Machine learning is shifting from being a tailor-made service of custom-crafted models into an era of productization. The new class of products and services coming out of this shift allows a diverse set of people to train their own models with their own data without the help of machine learning researchers. The tools are changing from being available only to large corporations that can afford it into the hands of smaller businesses that may be struggling to compete. As the reliance on specialists decreases for many use cases, it will become commonplace to see businesses of all sizes relying on artificial intelligence to improve some aspect of their daily operations. While the long-term impacts of this change have yet to be proven, the rise of AI automation tools has stirred all kinds of debate, especially around economics and jobs.

In terms of democratization, we still have a long way to go. However, we are making strides in increasing the accessibility of AI. This year may mark a milestone in the machine learning ecosystem. In January 2018, Gartner predicted that by 2019, organizations using self-service analytics and business intelligence tools would output more analytics than

© Leanne Luce 2019
L. Luce, *Artificial Intelligence for Fashion*,
https://doi.org/10.1007/978-1-4842-3931-5_12

professional analysts. While this statistic encompasses a wider breadth of technologies, AI plays a major role. The topic of the democratization of AI could not be more of the moment.

Software accessibility isn't the only driving force behind this shift. Access to data, the commoditization of specialized hardware, and an up-cropping of platforms to run that hardware are also playing their part.

Lowering the Barrier to Entry

It might not be obvious to professionals working outside the tech industry, but until recently, even experienced software developers might have had a difficult time implementing machine learning techniques. They might not have the background in mathematics or the domain knowledge to get machine learning models up and running. They also simply might not have had a reason to do it.

The progress that has been made in this field in the last five years has changed this dramatically. Every year, new tools are being introduced by the **open source** community, some of them with wide adoption. Products that are entering the market, especially **cloud services**, have also made it easier to train and run models.

Simplified Interfaces

As described in Chapter 7, there is a rise of increasingly powerful graphical interfaces for hosting data, training, and running models. With these GUIs, more can be done than ever before without writing code.

In terms of ease of use for the end user, solutions built for enterprise are built to have a more user-friendly GUI and automation in the back end, with the goal to increase the speed of execution. These platforms are built to make it easy for key stakeholders to understand and for scientists to implement.

One example of this is **DataRobot**, which offers enterprise products for running machine learning models. It's likely if you're using DataRobot that you have a data science team that can run and interpret experiments on the platform.

The goal of these platforms is to be able to do more faster, with little to no coding. It's not that these platforms make it easy for anyone at all to start using, but that for those who know the core concepts and how to use them, enterprise tools like DataRobot can provide powerful self-service automation.

Developer Tools

There are also tools accessible for developers looking to start implementing machine learning in the products they're working on.

Google's **Machine Learning Kit** (**ML Kit**) makes commonly used machine learning techniques available and production ready for developer use. They provide easy access to get started with these tools through their mobile and web app hosting service, Firebase.

ML Kit works with Google's **TensorFlow** Lite. TensorFlow is an open source software library for machine learning problems. It's meant to be able to run on everyday computers rather than just specialized machines. It also gives researchers a jump start at implementing neural networks. TensorFlow Lite is a version of this that is optimized for even lighter computation commonly required to run on mobile devices.

Access to Data

In March 2017, Google announced its acquisition of Kaggle, a data science platform mentioned in Chapter 7. Kaggle contains one of the largest repositories of datasets; an invaluable tool for learning and practicing machine learning. Before the acquisition, Google started releasing large labeled datasets to the platform, like a labeled set of YouTube videos. Access to this kind of resource was something that only large companies had before. Giving more people access to high-quality real-world data accelerates the rate of growth in the industry. For Google, this is a win, because when people can have better resources to learn from, they gain access to a larger pool of talent and help grow the market for their own services.

Kaggle isn't the only resource that people can go to for large datasets. Corporations like Amazon also provide lists of public datasets that are maintained by third parties and hosted on their platforms. In Amazon's case, this is through **Amazon Web Services** (**AWS**). Universities and government agencies frequently provide public data. New York–based startup Vigilant provides easy access to thousands of public records like these.

In September 2018, Google released Google Dataset Search, a tool to make it easy to search the Web for datasets: `https://toolbox.google.com/datasetsearch`.

Open Source

The word *open source* has been brought up a few times already in this chapter. Open source refers to software that has been made publicly available and can be modified or shared. Machine learning has progressed rapidly in part because of the open sourcing of tools and research.

Researchers have long been publishing academic papers containing their machine learning findings and the algorithms they used to achieve those results. What's different in the last five to ten years is that those papers are now linked to online media including videos, photos, repositories that host the code to run the algorithms, and even sometimes to demos that show you how they work or let you run them with your own data. Before, demos that ran neural networks and other complex models would not have been possible to host on the Web.

Specialized Hardware

Getting to where we are today with artificial intelligence has required the emergence of an entire ecosystem for machine learning to develop. Without computing power, AI processes like deep learning would not be possible. Until recently, the amount of computation required to run these models was more than a standard computer could handle.

GPUs and TPUs

Before the introduction of **graphics processing units (GPUs)** for machine learning in the early 2000s, neural networks were not feasible for production or even for research purposes. A GPU is similar to a **central processing unit (CPU)**, the electronic circuit that carries out instructions for your computer to work, but it can run many more computations at the same time than a CPU.

GPUs were originally designed for video games. In order to render graphics on a screen, video games run many complex calculations at once to quickly determine the value of every pixel on the screen based on what's happening in the game. Training neural networks and other machine learning models similarly requires a lot of computation. GPUs allow computers to run those computations in parallel, reducing the overall time needed to train. According to OpenAI, the amount of compute power used for training large jobs has doubled every 3.5 months since 2012.

Before 2012, it was very uncommon to use GPUs for machine learning. Between 2012 and 2014, infrastructure was built to make training on GPUs possible.

Around 2016, some of the approaches changed so that processes could run in parallel, and **tensor processing units (TPUs)** were introduced to the mix. The hardware has been constantly evolving to keep up with the demands for faster and more complex computation.

TPUs were introduced by Google. They are specialized hardware for deep learning, optimized for the basic operations required to train a neural network. The release of the TPU announcement garnered excitement in the industry because of a 15 to 30 times improvement in performance using these chips and even higher returns in terms of performance per watt. While these chips aren't available for purchase themselves, you can essentially rent them through Google's cloud computing services. Figure 12-1 shows a TPU circuit board and Google server center.

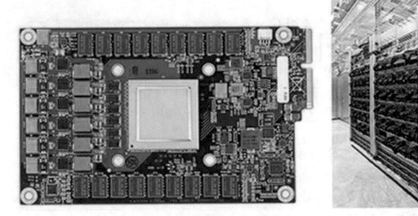

Figure 12-1.   *Tensor processing units (left) and their server center (right), image courtesy of Google.*

GPUs and TPUs aren't the full story. There have been a number of startups also building in the space. According to the *New York Times*, as of 2018, at least 45 startups were working on building machine-learning specialized chips alone.

Cloud Services

Even more recently, cloud computing using GPUs has become a viable solution for businesses looking to implement machine learning models without purchasing their own specialized hardware. While today, this is still expensive to maintain, new algorithms in development help optimize the use of expensive server time, reducing the costs significantly.

Cloud GPUs give experts the ability to run complex models on rented servers. FloydHub, Google Cloud GPUs, and Amazon ML all provide access to cloud computing of this type.

In January 2018, Google announced a new web service, **Cloud AutoML**, which allows businesses to use a number of their models without an in-house expert. Customers can complete tasks that are relevant to their businesses, like image classification, natural language processing, and language translation. Google plans to expand into other areas and offer a suite of easy-to-use machine learning tools.

Tutorials and Online Courses

In online education marketplaces like **Udemy** or **Coursera**, machine learning and related courses are increasing in prevalence. As someone with little technical experience, you can take one of these courses and get a step-by-step walk-through on machine learning. For example, you can learn how to build your own recommender system or train a neural network using popular programming languages.

Besides these marketplaces, tutorials in the space are easy to come across all over the Internet, on the blogs of startups and researchers.

Impact on Jobs

It will not be a world of man versus machine, it will be a world of man plus machine.

—Virginia Rometty, chair, president, and CEO, IBM

Artificial intelligence is often feared as a technology that will replace workers by automating their jobs. AI is good at automation and even outperforms humans at certain tasks, but not every task required to do most jobs. Throughout history, there have been more examples of automation creating jobs than taking them away.

The industrial revolution increased the amount of cloth a single weaver could produce by 50 times and decreased the labor per yard by 98%. This sounds like it would cause a near-elimination of workers in the industry, but in practice the inverse happened. With the reduction in the price of cloth came a steep increase in demand, creating four times more jobs.

Will this time be different? Technological change today is happening at a much quicker pace than it did in the 19th century. While the question remains impossible to answer with absolute confidence, many experts have weighed in with doubts that AI will be replacing entire jobs and industries anytime soon. However, some industries and the countries reliant on those industries will be affected more than others.

Ethics and the Future

Is artificial intelligence less than our intelligence?

—Spike Jonze, filmmaker

As the field of AI has matured, issues around its ethics have been called into question. Not only has it become more difficult to understand why a model makes a particular choice, but there has been little accountability in terms of understanding the biases inherent in the data they're trained on. This ambiguity has spurred a lot of criticism. How can we ensure that we aren't propagating our human biases when applying this new technology?

Race and Gender

A few alarming examples of AI biases about race and gender have sparked ethical conversations around discrimination. In early 2018, a study conducted by researcher Joy Buolamwini at the MIT Media Lab revealed starkly different error rates based on skin color and gender in facial recognition systems. In one of her examples, she shows that the accuracy of these systems was disproportionately more accurate for white men, with a 1% error rate in identifying gender, than it was for black women, who had up to 35% error rate.

At least in part, these biases lie in the data being passed through ML models to train them. As a potential step toward resolving these biases, Timnit Gebru and her coauthors from Microsoft Research proposed Datasheets for Datasets. The idea is that every dataset comes with a sort of nutrition label that explains details such as who made the dataset, where the data came from, and how it was created.

In all of computer science, it is estimated that women hold an estimated 13% of jobs. I personally believe that it will be difficult to resolve issues of bias without addressing issues of representation.

The Partnership on AI

The controversies around ethics expand beyond discrimination. Autonomous drone warfare and other war-related AI technologies also lie in a gray area. At what point do we begin introducing regulations around AI? Will regulation stunt progress in this field?

In January 2017, the Partnership on AI was formed by major technology companies to create open dialogue about ethics and implications of AI technology development. Since then, over 70 organizations and corporations worldwide have joined the nonprofit partnership. Their core tenants are around building technology that we understand.

The future really just depends on how we choose to use these powerful new tools.

Summary

The surge in new tools and hardware being introduced to the market has made artificial intelligence more accessible now than ever been before. Increases in compute power, open source machine learning models, specialized hardware, and cloud services specialized for the AI industry have all contributed to that growth.

The popularity and power of this technology has called a lot into question. Will it crash the job market? Is it ethical? We cannot predict what will happen in the future, but the technology itself will only do what the people controlling it ask it to do. We are a long way from AI taking on a mind of its own and destroying humanity; that discourse is a distraction from holding the humans who built it accountable.

Terminology from This Chapter

Amazon Web Services (AWS)—A suite of cloud computing services offered by Amazon on a paid subscription basis. See also *cloud services.*

Central processing units (CPUs)—A standard processor that exists in most consumer computers today. These are the "brains" of the computer, where operations are processed.

Cloud AutoML—A product offered by Google that makes it possible to train machine learning models with little machine learning expertise.

Cloud GPUs—Cloud computing services that specifically offer use of GPUs.

Cloud services—Broadly include services that are hosted over the Internet. Using cloud services, businesses or individuals don't need to maintain their own servers in order to offer or use certain services. In the context of machine learning, not needing to own a computer with a lot of processing power is very convenient.

Coursera –An online learning platform that offers not only courses in specialized topics, but also degrees. The platform works with top institutions like Yale and Stanford, as well as top companies like Google and IBM to offer high-quality courses.

DataRobot—A machine learning platform that automates some of the processes required to build and deploy machine learning models. It offers an easy-to-use GUI that makes it possible to implement these models without writing code.

Graphics processing units (GPUs)—A specialized electronic component often used in computers that require a lot of image processing or graphics rendering, as in video games. In the early 2000s, GPUs found their way into machine learning because of their higher processing power, which could be utilized in challenging computational problems.

Machine Learning Kit (ML Kit)—A product created by Google that makes it easy to implement certain machine learning products during application development.

Open source—Software in which the source code and the algorithms it is implementing are publicly accessible.

TensorFlow—Created by the Google Brain team at Google and released as open source in 2015. It is a software library primarily used for machine learning. TensorFlow Lite is an adaptation of TensorFlow that is built to run on Android devices, which have significantly less compute power than larger machines.

Tensor processing units (TPUs)—A hardware component introduced by Google to speed up computation during model training.

Udemy—An online education platform in which content generators can earn money in return for their courses. It is not a traditional academic program, and you cannot get college credit for taking courses there.

Bibliography

This bibliography is organized into 12 parts:

1. General References

2. Adversarial Examples

3. Chatbots, Virtual Style Assistants

4. Computer Vision, Visual Search

5. Data, Data Mining

6. Demand Forecasting

7. Ethics

8. Generative Models

9. Natural Language Processing

10. Neural Networks

11. Predictive Analytics, Recommendation Engines

12. Projects, Companies

General References

"Gartner Says Global Artificial Intelligence Business Value to Reach $1.2 Trillion in 2018." Gartner, 25 Apr. 2018, www.gartner.com/newsroom/id/3872933. Accessed Aug. 2018.

© Leanne Luce 2019
L. Luce, *Artificial Intelligence for Fashion*,
https://doi.org/10.1007/978-1-4842-3931-5

Abnett, Kate. "Is Fashion Ready for the AI Revolution?" The Business of Fashion, 7 Apr. 2016, www.businessoffashion.com/articles/fashion-tech/is-fashion-ready-for-the-ai-revolution. Accessed 22 Jan. 2018.

BOF Team. "BoF and Google Partner on Artificial Intelligence Experiment." The Business of Fashion, 30 Nov. 2017, www.businessoffashion.com/articles/sponsored-feature/bof-and-google-partner-on-artificial-intelligence-experiment. Accessed 14 Jan. 2018.

Dennis, Steve. "Many Unhappy Returns: E-Commerce's Achilles Heel." Forbes, 9 Aug. 2017, www.forbes.com/sites/stevendennis/2017/08/09/many-unhappy-returns-e-commerces-achilles-heel/#76296a344bf2. Accessed Feb. 2018.

Doupnik, Elizabeth. "Exclusive: How AI Predicts the Biggest Trends of the Season." WWD, 14 Mar. 2017, wwd.com/business-news/technology/ibm-watson-fashion-week-analysis-10842213/. Accessed Apr. 2018.

Gallo, Amy. "A Refresher on Regression Analysis." Harvard Business Review, 30 Nov. 2017, hbr.org/2015/11/a-refresher-on-regression-analysis. Accessed Mar. 2018.

Gerbert, Phillip, et al. "Putting Artificial Intelligence to Work." BCG, 28 Sept. 2017, www.bcg.com/en-us/publications/2017/technology-digital-strategy-putting-artificial-intelligence-work.aspx. Accessed Apr. 2018.

Knight, Will. "China's AI Awakening." MIT Technology Review, 10 Oct. 2017, www.technologyreview.com/s/609038/chinas-ai-awakening/. Accessed 16 April. 2018.

Luo, Ping, et al. "Large-Scale Fashion (DeepFashion) Database." The Chinese University of Hong Kong, Oct. 2016, mmlab.ie.cuhk.edu.hk/projects/DeepFashion.html. Accessed Apr. 2018.

Rasul, Kashif, and Han Xiao. "Fashion-MNIST." Zalando Research, research.zalando.com/welcome/mission/research-projects/fashion-mnist/. Accessed 11 Dec. 2017.

Xiao, et al. "Fashion-MNIST: a Novel Image Dataset for Benchmarking Machine Learning Algorithms." [Astro-Ph/0005112] A Determination of the Hubble Constant from Cepheid Distances and a Model of the Local Peculiar Velocity Field, 25 Aug. 2017, arxiv.org/abs/1708.07747v2. Accessed Aug. 2018.

Adversarial Examples

Athalye, et al. "Synthesizing Robust Adversarial Examples." [Astro-Ph/0005112] A Determination of the Hubble Constant from Cepheid Distances and a Model of the Local Peculiar Velocity Field, July 2017, arxiv.org/abs/1707.07397. Accessed May 2018.

Goodfellow, et al. "Explaining and Harnessing Adversarial Examples." Cornell University Library, 20 Dec. 2014, arxiv.org/abs/1412.6572v3. Accessed Apr. 2018.

Goodfellow, Ian, et al. "Attacking Machine Learning with Adversarial Examples." OpenAI Blog, 24 Feb. 2017, blog.openai.com/adversarial-example-research/. Accessed May 2018.

Chatbots, Virtual Style Assistants

"5 Ways Voice Assistance Is Reshaping Consumer Behavior." Think with Google, Aug. 2017, www.thinkwithgoogle.com/data-collections/voice-assistance-emerging-technologies/. Accessed Mar. 2018.

"Amazon Echo Look—Teardown." YouTube, 3 Nov. 2017, www.youtube.com/watch?v=nDRtwkKg8qU. Accessed July 2018.

"Messaging Apps Are Now Bigger than Social Networks." Business Insider, 20 Sept. 2016, www.businessinsider.com/the-messaging-app-report-2015-11. Accessed Aug. 2018.

Fumo, Nicola. "Rise of the AI Fashion Police." The Verge, 3 May 2017, www.theverge.com/2017/5/3/15522792/amazon-echo-look-alexa-style-assistant-ai-fashion. Accessed 15 Nov. 2017.

Garber, Megan. "When PARRY Met ELIZA: A Ridiculous Chatbot Conversation From 1972." The Atlantic, 9 June 2014, www.theatlantic.com/technology/archive/2014/06/when-parry-met-eliza-a-ridiculous-chatbot-conversation-from-1972/372428/. Accessed Aug. 2018.

Jacob, Neenu. "AI Could Become Your Personal Shopper." VentureBeat, 6 Oct. 2017, venturebeat.com/2017/09/24/ai-could-become-your-personal-shopper/. Accessed 2 Oct. 2018.

Kleinberg, Sara. "5 Ways Voice Assistance Is Reshaping Consumer Behavior." Think with Google, Jan. 2018, www.thinkwithgoogle.com/consumer-insights/voice-assistance-consumer-experience/. Accessed Jan. 2018.

Mau, Dhani. "How Brands and Startups Are Using AI to Help You Get Dressed." Fashionista, 17 Nov. 2017, fashionista.com/2017/11/fashion-brands-stylists-ai-artificial-intelligence-chatbots. Accessed 2 Oct. 2018.

McTear, Michael, et al. "Conversational Interfaces: Past and Present." SpringerLink, 1 Jan. 1970, link.springer.com/chapter/10.1007/978-3-319-32967-3_4. Accessed 2 Feb. 2018.

Pan, Jiaqi. "Conversational Interfaces: The Future of Chatbots." Chatbots Magazine, 25 Aug. 2017, chatbotsmagazine.com/conversational-interfaces-the-future-of-chatbots-18975a91fe5a. Accessed 4 Feb. 2018.

Rae, Haniya. "Inside Retail's Live Chat Revolution." Forbes 31 Mar. 2017, www.forbes.com/sites/haniyarae/2017/03/30/inside-retails-live-chat-revolution/#5099df982bce. Accessed 13 Apr. 2018.

Yao, Mariya. "4 Approaches To Natural Language Processing & Understanding." TOPBOTS, 29 July 2018, www.topbots.com/4-different-approaches-natural-language-processing-understanding/. Accessed 24 Feb. 2018.

Computer Vision, Visual Search

Edvinsson, Johan. "Machine Learning at Condé Nast, Part 2: Handbag Brand and Color Detection." Condé Nast Technology, 6 Nov. 2017, technology.condenast.com/story/handbag-brand-and-color-detection. Accessed July 2018.

Fergus, Rob, et al. "Tiny Images Dataset." New York University, horatio.cs.nyu.edu/mit/tiny/data/index.html. Accessed Apr. 2018.

Karpathy, Andrej, and Justin Johnson. "CS231n Convolutional Neural Networks for Visual Recognition." Stanford University, cs231n.github.io/classification/. Accessed Apr. 2018.

Li, Fei-Fei, et al. "Spatial Localization and Detection." Stanford University, 1 Feb. 20161, cs231n.stanford.edu/slides/2016/winter1516_lecture8.pdf. Accessed Apr. 2018.

Lusch, David P. "Digital Image Classification." Michigan State University, Oct. 2015, lees.geo.msu.edu/courses/geo827/lecture_10_classification.pdf. Accessed Mar. 2018.

Mallick, Satya. "Image Recognition and Object Detection: Part 1." Learn OpenCV, 14 Nov. 2016, www.learnopencv.com/image-recognition-and-object-detection-part1/. Accessed Apr. 2018.

Manoff, Jill. "To Find It, Just Boohoo It." Glossy Magazine, 27 Feb. 2018, www.glossy.co/ecommerce/to-find-it-just-boohoo-it-how-the-fast-fashion-retailer-is-making-a-go-of-visual-search. Accessed Mar. 2018.

Urtasun, Raquel. "Computer Vision: Filtering." University of Toronto, 10 Jan. 2013, www.cs.toronto.edu/~urtasun/courses/CV/lecture02.pdf. Accessed Mar. 2018.

Data, Data Mining

"What Is a Data Warehouse?." Amazon Web Services, aws.amazon.com/data-warehouse/. Accessed May 2018.

Emerging Technology from the arXiv. "AI Reveals Global Clothing Preferences by Data-Mining Instagram Photos." MIT Technology Review, 15 June 2017, www.technologyreview.com/s/608116/data-mining-100-million-instagram-photos-reveals-global-clothing-patterns/. Accessed Apr. 2018.

Henke, Nicolaus, et al. "The Age of Analytics: Competing in a Data-Driven World." McKinsey & Company, Dec. 2016, www.mckinsey.com/~/media/McKinsey/Business%20Functions/McKinsey%20Analytics/Our%20Insights/The%20age%20of%20analytics%20Competing%20in%20a%20data%20driven%20world/MGI-The-Age-of-Analytics-Full-report.ashx. Accessed Mar. 2018.

Lee, Doris Jung-Lin, et al. "Identifying Fashion Accounts in Social Networks." KDD Fashion 2017, University of Illinois, Urbana-Champaign, Aug. 2017, kddfashion2017.mybluemix.net/final_submissions/ML4Fashion_paper_21.pdf. Accessed Apr. 2018.

Demand Forecasting

"Autoregressive Integrated Moving Average—ARIMA." Investopedia, www.investopedia.com/terms/a/autoregressive-integrated-moving-average-arima.asp. Accessed 2 Aug. 2018.

de las Heras Torres, Roman Josue. "7 Ways Time-Series Forecasting Differs from Machine Learning." Oracle + DataScience.com, 29 May 2018, www.datascience.com/blog/time-series-forecasting-machine-learning-differences. Accessed Aug. 2018.

Taylor, Sean J, and Benjamin Letham. "Forecasting at Scale." PeerJ, 27 Sept. 2017, peerj.com/preprints/3190/. Accessed Aug. 2018.

Ethics

Crawford, Kate. "Opinion | Artificial Intelligence's White Guy Problem." The New York Times, 20 Jan. 2018, www.nytimes.com/2016/06/26/opinion/sunday/artificial-intelligences-white-guy-problem.html?_r=1. Accessed Aug. 2018.

Hardesty, Larry, and MIT News Office. "Study Finds Gender and Skin-Type Bias in Commercial Artificial-Intelligence Systems." MIT News, 11 Feb. 2018, news.mit.edu/2018/study-finds-gender-skin-type-bias-artificial-intelligence-systems-0212. Accessed Aug. 2018.

Nickelsburg, Monica. "Why Is AI Female? How Our Ideas about Sex and Service Influence the Personalities We Give Machines." GeekWire, 27 Nov. 2017, www.geekwire.com/2016/why-is-ai-female-how-our-ideas-about-sex-and-service-influence-the-personalities-we-give-machines/. Accessed Aug. 2018.

Watercutter, Angela. "Ex Machina Has a Serious Fembot Problem." Wired, 6 June 2017, www.wired.com/2015/04/ex-machina-turing-bechdel-test/. Accessed Aug. 2018.

Generative Models

Deverall, Jaime, et al. "Using Generative Adversarial Networks to Design Shoes: The Preliminary Steps." Stanford University, 13 June 2017, cs231n.stanford.edu/reports/2017/pdfs/119.pdf. Accessed 2 May 2018.

Doersch, Carl. "Tutorial on Variational Autoencoders." [Astro-Ph/0005112] A Determination of the Hubble Constant from Cepheid Distances and a Model of the Local Peculiar Velocity Field, 13 June 2016, arxiv.org/abs/1606.05908. Accessed June 2018.

Floydhub. "Floydhub/Dcgan." GitHub, github.com/floydhub/dcgan. Accessed Apr. 2018.

Ganesan, et al. "Fashioning with Networks: Neural Style Transfer to Design Clothes." [Astro-Ph/0005112] A Determination of the Hubble Constant from Cepheid Distances and a Model of the Local Peculiar Velocity Field, 31 July 2017, arxiv.org/abs/1707.09899. Accessed 5 Feb. 2018.

Goodfellow, et al. "Generative Adversarial Networks." [Astro-Ph/0005112] A Determination of the Hubble Constant from Cepheid Distances and a Model of the Local Peculiar Velocity Field, 10 June 2014, arxiv.org/abs/1406.2661. Accessed Apr. 2018.

Heinrich, Greg. "Photo Editing with Generative Adversarial Networks (Part 1)." NVIDIA Developer Blog, 20 Apr. 2017, devblogs.nvidia.com/photo-editing-generative-adversarial-networks-1/. Accessed July 2018.

Jetchev, et al. "The Conditional Analogy GAN: Swapping Fashion Articles on People Images." [Astro-Ph/0005112] A Determination of the Hubble Constant from Cepheid Distances and a Model of the Local Peculiar Velocity Field, 14 Sept. 2017, arxiv.org/abs/1709.04695v1. Accessed May 2018.

Karpathy, Andrej, et al. "Generative Models." OpenAI Blog, 16 June 2016, blog.openai.com/generative-models/. Accessed May 2018.

Karras, Tero, et al. "Progressive Growing of GANs for Improved Quality, Stability, and Variation." Nvidia Research, 30 Apr. 2018, research.nvidia.com/publication/2017-10_Progressive-Growing-of. Accessed May 2018.

Kato, Natsumi, et al. "Crowd Sourcing Clothes Design Directed by Adversarial Neural Networks." NIPS 2017, Yoichi Ochiai, University of Tsukuba, 2017, nips2017creativity.github.io/doc/Crowd_Sourcing_Clothes_Design.pdf. Accessed 28 Jan. 2018.

Knight, Will. "Amazon Has Developed an AI Fashion Designer." MIT Technology Review, 6 Sept. 2017, www.technologyreview.com/s/608668/amazon-has-developed-an-ai-fashion-designer/. Accessed Mar. 2018.

Knight, Will. "This Inventor Applied Game Theory to Machine Learning to Make Computers Smarter." MIT Technology Review, 22 Aug. 2017, www.technologyreview.com/lists/innovators-under-35/2017/inventor/ian-goodfellow/. Accessed 2 Oct. 2018.

Mau, Dhani. "The 10 Biggest U.S. Apparel Companies." Fashionista, 2 July 2015, fashionista.com/2015/07/most-valuable-american-brands. Accessed 2 July 2018.

Monn, Dominic. "Deep Convolutional Generative Adversarial Networks with TensorFlow." O'Reilly, 2 Nov. 2017, www.oreilly.com/ideas/deep-convolutional-generative-adversarial-networks-with-tensorflow. Accessed June 2018.

Pandey, Prakash. "Deep Generative Models." Towards Data Science, 31 Jan. 2018, towardsdatascience.com/deep-generative-models-25ab2821afd3. Accessed Mar. 2018.

Phillip, et al. "Image-to-Image Translation with Conditional Adversarial Networks." [Astro-Ph/0005112] A Determination of the Hubble Constant from Cepheid Distances and a Model of the Local Peculiar Velocity Field, 22 Nov. 2017, arxiv.org/abs/1611.07004. Accessed July 2018.

Sbai, Othman, et al. "DeSIGN: Design Inspiration from Generative Networks." GroundAI, Apr. 2018, www.groundai.com/project/design-design-inspiration-from-generative-networks/. Accessed May 2018.

Shafkat, Irhum. "Intuitively Understanding Variational Autoencoders." Towards Data Science, 4 Feb. 2018, towardsdatascience.com/intuitively-understanding-variational-autoencoders-1bfe67eb5daf. Accessed Apr. 2018.

Wang, et al. "Generative Image Modeling Using Style and Structure Adversarial Networks." [Astro-Ph/0005112] A Determination of the Hubble Constant from Cepheid Distances and a Model of the Local Peculiar Velocity Field, 26 Mar. 2016, arxiv.org/abs/1603.05631v2. Accessed May 2018.

Weston, Phoebe. "AI 'Stylist' Learns Your Fashion, Invents Your Next Outfit." Daily Mail Online, 16 Nov. 2017, www.dailymail.co.uk/sciencetech/article-5089505/AI-stylist-learns-fashion-invents-outfit.html. Accessed 13 Dec. 2017.

BIBLIOGRAPHY

Natural Language Processing

"Recursive Deep Models for Semantic Compositionality Over a Sentiment Treebank." Stanford University: NLP, nlp.stanford.edu:8080/sentiment/rntnDemo.html. Accessed 14 Apr. 2018.

"Vector Representations of Words." TensorFlow, www.tensorflow.org/tutorials/representation/word2vec#the_skip_gram_model. Accessed May 2018.

Dalinina, Ruslana. "Word Embeddings: An NLP Crash Course." Oracle + DataScience.com, 10 Oct. 2017, www.datascience.com/resources/notebooks/word-embeddings-in-python. Accessed Apr. 2018.

Gaskin, Jennifer L., et al. "An Introduction to Recommendation Engines." Dataconomy, 30 May 2016, dataconomy.com/2015/03/an-introduction-to-recommendation-engines/. Accessed May 2018.

Jurafsky, Daniel, and James H. Martin. "Sequence Processing with Recurrent Networks" and "Formal Grammars of English" *Speech and Language Processing*. Pearson, 2014. Chapter 9-10. Print.

MacCartney, Bill. "Understanding Natural Language Understanding." Stanford University, ACM SIGAI, 16 July 2014, nlp.stanford.edu/~wcmac/papers/20140716-UNLU.pdf. Accessed 25 Mar. 2018.

Manning, Christopher D., et al. "Tokenization" *Introduction to Information Retrieval*. Cambridge University Press, 2017. Print.

McCormick, Chris. "Word2Vec Tutorial—The Skip-Gram Model." Chris McCormick—Machine Learning Tutorials, 19 Apr. 2016, mccormickml.com/2016/04/19/word2vec-tutorial-the-skip-gram-model/. Accessed June 2018.

Nichols, Nate. "Natural Language Processing and Natural Language Generation: What's the Difference?" Narrative Science, 24 Apr. 2017, narrativescience.com/Resources/Resource-Library/Article-Detail-Page/natural-language-processing-and-natural-language-generation-whats-the-difference. Accessed 23 Mar. 2018.

Orr, Dave. "50,000 Lessons on How to Read: a Relation Extraction Corpus." Google AI Blog, 11 Apr. 2013, `ai.googleblog.com/2013/04/50000-lessons-on-how-to-read-relation.html`. Accessed 27 Apr. 2018.

Socher, Richard, et al. "Recursive Deep Models for Semantic Compositionality Over a Sentiment Treebank." Stanford University, `nlp.stanford.edu/~socherr/EMNLP2013_RNTN.pdf`.

Neural Networks

Dancho, Matt. "Time Series Analysis: KERAS LSTM Deep Learning—Part 1." Business Science, 18 Apr. 2018, `www.business-science.io/timeseries-analysis/2018/04/18/keras-lstm-sunspots-time-series-prediction.html`. Accessed Aug. 2018.

Le, Quoc, and Barret Zoph. "Using Machine Learning to Explore Neural Network Architecture." Google AI Blog 17 May 2017, `ai.googleblog.com/2017/05/using-machine-learning-to-explore.html`. Accessed Apr. 2018.

Leverington, David. "A Basic Introduction to Feedforward Backpropagation Neural Networks." Texas Tech University, 2009, `www.webpages.ttu.edu/dleverin/neural_network/neural_networks.html`. Accessed Apr. 2018.

Maini, Vishal. "Machine Learning for Humans, Part 4: Neural Networks & Deep Learning." Machine Learning for Humans, 19 Aug. 2017, `medium.com/machine-learning-for-humans/neural-networks-deep-learning-cdad8aeae49b`. Accessed 12 Mar. 2018.

Olah, Christopher. "Understanding LSTM Networks." Colah's Blog, 27 Aug. 2015, `colah.github.io/posts/2015-08-Understanding-LSTMs/`. Accessed June 2018.

Szegedy, Christian, et al. "Going Deeper with Convolutions." CVPR 2015, static.googleusercontent.com/media/research.google.com/en//pubs/archive/43022.pdf. Accessed Mar. 2018.

Zoph, Barret, and Quoc V. Le. "Neural Architecture Search with Reinforcement Learning." Google AI, ICLR 2017, ai.google/research/pubs/pub45826. Accessed Aug. 2018.

Predictive Analytics, Recommendation Engines

"Fit Analytics Case Study with Google Cloud." Google Cloud Blog, cloud.google.com/customers/fit-analytics/. Accessed Aug. 2018.

"What Are Product Recommendation Engines? And the Various Versions of Them?" Towards Data Science, 28 Sept. 2017, towardsdatascience.com/what-are-product-recommendation-engines-and-the-various-versions-of-them-9dcab4ee26d5. Accessed June 2018.

Ariker, Matt, et al. "Personalizing at Scale." McKinsey & Company, Nov. 2015, www.mckinsey.com/business-functions/marketing-and-sales/our-insights/personalizing-at-scale. Accessed 2 Oct. 2018.

Clark, Jesse. "The Curious Connection Between Warehouse Maps, Movie Recommendations, and Structural Biology." Multithreaded | Stitch Fix Technology Blog, 31 Aug. 2017, multithreaded.stitchfix.com/blog/2017/08/31/warehouse-layouts/. Accessed May 2018.

Colson, Eric. "Personalizing Beyond the Point of No Return." Multithreaded | Stitch Fix Technology Blog, 7 July 2015, multithreaded.stitchfix.com/blog/2015/07/07/personalizing-beyond-the-point-of-no-return/. Accessed May 2018.

Colson, Eric, et al. "Stitch Fix Algorithms Tour." Stitch Fix, algorithms-tour.stitchfix.com/#data-platform. Accessed 20 Feb. 2018.

Foley, Patrick, and John McDonnell. "What the SATs Taught Us About Finding the Perfect Fit." Multithreaded | Stitch Fix Technology Blog, 13 Dec. 2017, multithreaded.stitchfix.com/blog/2017/12/13/latentsize/. Accessed May 2018.

Gaskin, Jennifer L., et al. "An Introduction to Recommendation Engines." Dataconomy, 30 May 2016, dataconomy.com/2015/03/an-introduction-to-recommendation-engines/. Accessed May 2018.

Ivens, et al. "The Use of Machine Learning Algorithms in Recommender Systems: A Systematic Review." [Astro-Ph/0005112] A Determination of the Hubble Constant from Cepheid Distances and a Model of the Local Peculiar Velocity Field, 24 Feb. 2016, arxiv.org/abs/1511.05263. Accessed May 2018.

Mankar, Ashwini N., and Gogate Uttara Dhananjay. "A Review of Different Techniques for Recommender Systems." IJSRD || National Conference on Technological Advancement and Automatization in Engineering, Jan. 2016, www.ijsrd.com/articles/NCTAAP157.pdf. Accessed May 2018.

Moody, Chris. "Stop Using word2vec." Multithreaded | Stitch Fix Technology Blog, 18 Oct. 2017, multithreaded.stitchfix.com/blog/2017/10/18/stop-using-word2vec/. Accessed May 2018.

Moody, Chris. "Word Tensors." Multithreaded | Stitch Fix Technology Blog, 25 Oct. 2017, multithreaded.stitchfix.com/blog/2017/10/25/word-tensors/. Accessed May 2018.

Naik, Deepa. "Understanding Recommendation Engines in AI." Humans for AI, Medium, 3 June 2017, medium.com/@humansforai/recommendation-engines-e431b6b6b446. Accessed Apr. 2018.

Schneider, Anna, and Alex Smolyanskaya. "Lumpers and Splitters: Tensions in Taxonomies." Multithreaded | Stitch Fix Technology Blog, 5 Apr. 2018, multithreaded.stitchfix.com/blog/2018/04/05/lumpers-and-splitters/. Accessed May 2018.

Robotics, Impact

"Automation and Anxiety." The Economist, 25 June 2016, www.economist.com/special-report/2016/06/25/automation-and-anxiety. Accessed Mar. 2018.

Dobbs, Richard, et al. "The Four Global Forces Breaking All the Trends." McKinsey & Company, Apr. 2015, www.mckinsey.com/business-functions/strategy-and-corporate-finance/our-insights/the-four-global-forces-breaking-all-the-trends. Accessed July 2018.

Futch, Mike. "Rise of the Warehouse Robots." Material Handling and Logistics (MHL News), 18 Oct. 2017, www.mhlnews.com/technology-automation/rise-warehouse-robots. Accessed Aug. 2018.

Grier, William. "A Historical Timeline Perspective to the Crash of Apparel Retail." LinkedIn, 7 Aug. 2018, www.linkedin.com/pulse/historical-timeline-perspective-crash-apparel-retail-william-grier/?lipi=urn%3Ali%3Apage%3Ad_flagship3_feed%3BkMi%2BG318TI6BpIBq3p1bCA%3D%3D. Accessed Aug. 2018.

Lohr, Steve. "'The Beginning of a Wave': A.I. Tiptoes into the Workplace." The New York Times, 5 Aug. 2018, www.nytimes.com/2018/08/05/technology/workplace-ai.html. Accessed Aug. 2018.

Simon, Matt. "What Is a Robot?" Wired, 24 Aug. 2017, www.wired.com/story/what-is-a-robot/. Accessed July 2018.

Spinney, Laura. "Can Robots Make Art?" Nature, 27 Apr. 2018, www.nature.com/articles/d41586-018-04989-2. Accessed 2 Oct. 2018.

Specialized Hardware

"New Specialized AI Chips May Revolutionize the $1.6-Plus Trillion Market Cap Semiconductor Industry." 13D Research, 1 July 2018, latest.13d.com/specialized-ai-chips-revolutionize-semiconductor-industry-tech-6b825d292282. Accessed Aug. 2018.

Amodei, Dario, and Danny Hernandez. "AI and Compute." OpenAI Blog, 16 May 2018, blog.openai.com/ai-and-compute/. Accessed June 2018.

Metz, Cade. "Big Bets on A.I. Open a New Frontier for Chip Start-Ups, Too." The New York Times, 14 Jan. 2018, www.nytimes.com/2018/01/14/technology/artificial-intelligence-chip-start-ups.html. Accessed June 2018.

Sato, Kaz, et al. "An in-Depth Look at Google's First Tensor Processing Unit (TPU)." Google Cloud Blog, 12 May 2017, cloud.google.com/blog/products/gcp/an-in-depth-look-at-googles-first-tensor-processing-unit-tpu. Accessed June 2018.

Projects, Companies

DataRobot (Enterprise AI), www.datarobot.com

Edited (Fashion Analytics), https://edited.com/

Facebook, Prophet (Forecasting Model), https://facebook.github.io/prophet/

Floydhub (Cloud ML), www.floydhub.com

Google Cloud AutoML, https://cloud.google.com/automl/

Google Dataset Search, https://toolbox.google.com/datasetsearch

IBM Cognitive Fashion, https://cognitivefashion.github.io/

MemoMi (Smart Mirrors), http://memorymirror.com/

MLJAR (Cloud ML), https://mljar.com/

Mode AI (Conversational Shopping), http://mode.ai/

SoftWear Automation (Robotic Sewing), http://softwearautomation.com/

TensorFlow (Open Source ML), www.tensorflow.org

Tesserai (Production-Quality AI), https://tesserai.com

Trendage AI (Complete the Look), www.trendage.com

Trendalytics (Fashion Analytics), www.trendalytics.co

Index

A

Adaption, 65
Adobe Photoshop, 46
Adversarial additions, 69
Adversarial image overlays, 68–69
Adversarial objects, 70
Adversarial patch, 69
Aggregating trends, 132
AI fashion blogger, 132–133
AI fashion designer, 125
   artificial creativity, 126–127
   garment, 127–128
   turning sketches, 129
Alt text, 56
Amazon Echo Look, 79–80
Amazon Web Services (AWS), 188
Application programming
   interfaces (APIs), 29, 145
Applied AI/narrow AI, 83
Articulated robots, 171
Artificial general intelligence
   (AGI), 83
Artificial intelligence (AI)
   biases, 193
   dangers, 84–85
   jobs, 192
   partnership, 193

   pitfalls, 84
   tools and techniques, 13–14
Automatic speech recognition
   (ASR), 77
Automation, 176–177
Autoregressive integrated moving
   average (ARIMA), 163
Autoregressive models, 130

B

Back-end database, 145
Backpropagation, 64
Bipedal, 167, 179
Bot-to-Bot interaction, 27
Brand subscriptions, 92

C

Central processing unit
   (CPU), 189
Chatbots, 23
Cleaning data, 113
Cloud AutoML, 191
Cloud GPUs, 191
Collaborative filtering, 95
   item-based approach, 96
   user-based approach, 96

© Leanne Luce 2019
L. Luce, *Artificial Intelligence for Fashion*,
https://doi.org/10.1007/978-1-4842-3931-5

Printed in the United States
By Bookmasters